CRC SERIES IN AGRICULTURE

Editor-in-Chief

Angus A. Hanson, Ph.D.
Vice President-Research
W-L Research, Inc.
Highland, Maryland

HANDBOOK OF SOILS AND CLIMATE IN AGRICULTURE

Editor
Victor J. Kilmer (Deceased)
Chief
Soils and Fertilizer Research Branch
National Fertilizer Development Center
Tennessee Valley Authority
Muscle Shoals, Alabama

HANDBOOK OF PLANT SCIENCE IN AGRICULTURE

Editor
B. R. Christie, Ph.D.
Professor
Department of Crop Science
Ontario Agricultural College
University of Guelph
Guelph, Ontario, Canada

HANDBOOK OF PEST MANAGEMENT IN AGRICULTURE

Editor
David Pimentel, Ph.D.
Professor
Department of Entomology
New York College of Agricultural
and Life Sciences
Cornell University
Ithaca, New York

HANDBOOK OF ENGINEERING IN AGRICULTURE

Editor
R. H. Brown, Ph.D.
Chairman
Division of Agricultural Engineering
Agricultural Engineering Center
University of Georgia
Athens, Georgia

HANDBOOK OF TRANSPORTATION AND MARKETING IN AGRICULTURE

Editor
Essex E. Finney, Jr., Ph.D.
Assistant Center Director
Agricultural Research Center
U.S. Department of Agriculture
Beltsville, Maryland

HANDBOOK OF PROCESSING AND UTILIZATION IN AGRICULTURE

Editor
Ivan A. Wolff, Ph.D. (Retired)
Director
Eastern Regional Research Center
Science and Education Administration
U.S. Department of Agriculture
Philadelphia, Pennsylvania

CRC Handbook

Engineering in Agriculture

Volume I
Crop Production Engineering

Editor

R. H. Brown, P.E.

Chairman Emeritus
Division of Agricultural Engineering
University of Georgia
Athens, Georgia

CRC Series in Agriculture
A. A. Hanson, Editor-in-Chief
Vice President-Research
W-L Research, Inc.
Highland, Maryland

CRC Press, Inc.
Boca Raton, Florida

Library of Congress Cataloging-in-Publication Data

CRC handbook of engineering in agriculture.

  (CRC series in agriculture)

  Includes bibliographies and indexes.

  1. Agricultural engineering—Handbooks, manuals,
etc.   I. Brown, R. H. (Robert H.)   II. Series.
S675.C73 1988   630   87-21870
ISBN 0-8493-3861-1 (v. 1)
ISBN 0-8493-3862-X (v. 2)
ISBN 0-8493-3863-8 (v. 3)

© 1988 by CRC Press, Inc.

International Standard Book Number 0-8493-3861-1 (v. 1)
International Standard Book Number 0-8493-3862-X (v. 2)
International Standard Book Number 0-8493-3863-8 (v. 3)

Library of Congress Card Number 87-21870
Printed in the United States

EDITOR-IN-CHIEF

Angus A. Hanson, Ph.D., is Vice President-Research, W-L Research, Inc., Highland, Maryland, and has had broad experience in agricultural research and development. He is a graduate of the University of British Columbia, Vancouver, and McGill University, Quebec, and received the Ph.D. degree from the Pennsylvania State University, University Park, in 1951.

An employee of the U.S. Department of Agriculture from 1949 to 1979, Dr. Hanson worked as a Research Geneticist at University Park, Pa., 1949 to 1952, and at Beltsville, Md., serving successively as Research Leader for Grass and Turf Investigations, 1953 to 1965, Chief of the Forage and Range Research Branch, 1965 to 1972, and Director of the Beltsville Agricultural Research Center, 1972 to 1979. He has been appointed to a number of national and regional task forces charged with assessing research needs and priorities, and has participated in reviewing agricultural needs and research programs in various foreign countries. As Director at Beltsville, he was directly responsible for programs that included most dimensions of agricultural research.

In his personal research, he has emphasized the improvement of forage crops, breeding and management of turfgrasses, and the breeding of alfalfa for multiple pest resistance, persistence, quality, and sustained yield. He is the author of over 100 technical and popular articles on forage crops and turfgrasses, and has served as Editor of *Crop Science* and the *Journal of Environmental Quality*.

PREFACE

CRC SERIES IN AGRICULTURE

Agriculture, because of its pivotal role in the development of civilized societies, contributed much to the development of various scientific disciplines. Thus, agricultural pursuits led to the practical application of chemistry, and gave rise to such major disciplines as economics and statistics. The expansion of scientific frontiers, and the concomitant specilization within disciplines, has proceeded to the point where agricultural scientists classify themselves in an array of disciplines and subdisciplines, i.e., nematologist, geneticist, physicist, virologist, and so forth. Nevertheless, within the framework of these various disciplines and mission oriented agricultural research, information of primary interest and concern in the solution of agriculturally oriented problems is generated. Although some of the basic information finds its way into the plethora of reference books available within most disciplines, no attempt has been made to develop a comprehensive handbook series for the agricultural sciences.

It is recognized that there are serious difficulties in developing a meaningful handbook series in agriculture because of the range and complexity of agricultural enterprises. In fact, the single common denominator that applies to all agricultural scientists is their universal concern with at least some aspect of the production and utilization of farm products. The disciplines and resources that are called for in a specific investigation are either the same or similar to those utilized in any area of biological research, or in any one of several fields of scientific endeavor.

The sections in this handbook series reflect the input of different editors and advisory boards, and as a consequence, there is considerable variation in both the depth and coverage offered within a given area. However, an attempt has been made throughout to bring together pertinent information that will serve the needs of nonspecialists, provide a quick reference to material that might otherwise be difficult to locate, and furnish a starting point for further study.

The project was undertaken with the realization that the initial volumes in the series could have some obvious deficiencies that will necessitate subsequent revisions. In the meantime, it is felt that the primary objectives of the Section Editors and their Advisory Boards has been met in this first edition.

A. A. Hanson
Editor-in-Chief

ADVISORY BOARD

CONTRIBUTORS

B. J. Barfield, Ph.D.
Department of Agricultural Engineering
University of Kentucky
Lexington, Kentucky

D. B. Brooker, Ph.D.
Professor Emeritus
Department of Agricultural Engineering
University of Missouri
Columbia, Missouri

R. H. Brown, P.E., Ph.D.
Chairman Emeritus
Division of Agricultural Engineering
University of Georgia
Athens, Georgia

R. R. Bruce, Ph.D.
Soil Scientist
Southern Piedmont Research Center
USDA
Watkinsville, Georgia

J. L. Butler, Ph.D.
Research Leader
Crop Systems Research Unit
USDA
Tifton, Georgia

W. J. Chancellor, Ph.D.
Professor
Department of Agricultural Engineering
University of California
Davis, California

J. L. Chesness, Ph.D.
Professor
Department of Agricultural Engineering
University of Georgia
Athens, Georgia

C. M. Christensen, Ph.D.
Professor Emeritus
Department of Plant Pathology
University of Minnesota
St. Paul, Minnesota

D. S. Chung, Ph.D.
Professor
Department of Agricultural Engineering
Kansas State University
Manhattan, Kansas

C. J. W. Drablos, Ph.D.
Professor
Department of Agricultural Engineering
University of Illinois
Urbana, Illinois

W. C. Fairbank
Extension Agricultural Engineer
Department of Soil and Environmental
Science
University of California
Riverside, California

P. R. Goodrich, Ph.D.
Associate Professor
Department of Agricultural Engineering
University of Minnesota
St. Paul, Minnesota

J. W. Goodrum, Ph.D
Associate Professor
Department of Agricultural Engineering
University of Georgia
Athens, Georgia

W. C. Hammond, Ph.D.
Head
Department of Extension Engineering
Cooperative Extension Services
Athens, Georgia

P. K. Harein, Ph.D.
Department of Entomology, Fisheries, &
Wildlife
University of Minnesota
St. Paul, Minnesota

J. C. Hayes, Ph.D.
Department of Agricultural Engineering
Clemson University
Clemson, South Carolina

J. G. Hendrick
Agricultural Engineer
National Tillage Machinery Lab
USDA
Auburn, Alabama

T. A. Howell, Ph.D.
Agricultural Engineer
Conservation and Production Research Lab
USDA
Bushland, Texas

R. W. Irwin, Ph.D.
Professor
School of Engineering
University of Guelph
Guelph, Ontario, Canada

F. C. Ives, P.E.
Design Engineer
Soil Conservation Service
USDA
Champaign, Illinois

J. M. Laflen, Ph.D.
Research Leader
National Soil Erosion Research Lab
USDA
West Lafayette, Indiana

J. H. Lehr, Ph.D.
Executive Director
National Water Well Association
Dublin, Ohio

W. D. Lembke, Ph.D.
Professor Emeritus
Department of Agricultural Engineering
University of Illinois
Urbana, Illinois

L. Lyles, Ph.D.
Research Leader
Wind Erosion Unit
USDA
Manhattan, Kansas

H. B. Manbeck, Ph.D.
Professor
Department of Agricultural Engineering
Pennsylvania State University
Univerisity Park, Pennsylvania

J. R. Miner, Ph.D.
Associate Director and Professor
Department of International Research and
Development
Oregon State University
Corvallis, Oregon

C. H. Moss, P.E.
Manager Civil Engineer
T. E. Stivers Organization, Inc.
Decatur, Georgia

B. H. Nolte, Ph. D.
Professor
Department of Agricultural Engineering
Ohio State University
Columbus, Ohio

J. C. Nye
Professor
Department of Agricultural Engineering
Louisiana State University
Baton Rouge, Louisiana

C. H. Pair
Engineer
USDA
Boise, Idaho

L. K. Pickett, Ph.D.
Senior Project Engineer
Department of Advanced Engineering
Case International
Hinsdale, Illinois

J. H. Poehlman
Largo, Florida

L. M. Safley, Jr., Ph.D.
Assistant Professor
Department of Agricultural Engineering
University of Tennessee
Knoxville, Tennessee

G. O. Schwab, Ph.D.
Professor Emeritus
Department of Agricultural Engineering
Ohio State University
Columbus, Ohio

Hollis Shull
Engineer
USDA
University of Nebraska
Lincoln, Nebraska

J. W. Simons
Research Associate
Department of Agricultural Engineering
University of Georgia
Athens, Georgia

R. P. Singh, Ph.D.
Professor
Department of Agricultural Engineering
University of California
Davis, California

R. E. Sneed, Ph.D.
Professor
Department of Biological and Agricultural
Engineering
North Carolina State University
Raleigh, North Carolina

J. M. Steichen, Ph.D.
Associate Professor
Department of Agricultural Engineering
Kansas State University
Manhattan, Kansas

C. W. Suggs, Ph.D.
Professor
Department of Biological and Agricultural
Engineering
North Carolina State University
Raleigh, North Carolina

J. M. Sweeten, Ph.D.
Extension Agricultural Engineer
Texas Agricultural Extension Service
Texas A&M University
College Station, Texas

J. R. Talbot
National Soil Engineer
USDA
Washington, D. C.

E. D. Threadgill, Ph.D.
Chairman
Department of Agricultural Engineering
University of Georgia
Athens, Georgia

D. H. Vanderholm
Associate Dean
Institute of Agriculture and Natural Resources
University of Nebraska
Lincoln, Nebraska

G. L. Van Wicklen, Ph.D.
Assistant Professor
Department of Agricultural Engineering
University of Georgia
Athens, Georgia

N. L. West
Project Engineer
John Deere Harvester
East Moline, Illinois

I. L. Winsett
Sales Engineer
Ronk Electrical Indusrties
Nokomis, Illinois

TABLE OF CONTENTS

Volume II

EROSION CONTROL ENGINEERING

DRAINAGE ENGINEERING

Volume III

PHYSIOLOGICAL PARAMETERS AND REQUIREMENTS OF LIVESTOCK POULTRY

STRUCTURAL DESIGNS, REQUIREMENTS, AND SYSTEMS

ELECTRICAL SYSTEMS AND APPLIANCES FOR AGRICULTURAL STRUCTURES

FEED AND CROP STORAGES

APPENDIX

Engineering Properties of Agricultural Soils

SOIL CLASSIFICATION BY PARTICLE SIZE — TEXTURAL TRIANGLE

James R. Talbot

SOIL CLASSIFICATION

Soil behavior is influenced by many factors relating to the physical and chemical make-up of a soil and the condition in which a soil exists. These factors, to some degree, interact or influence each other. Physical behavior of soil is influenced by the chemical composition and its interaction with the distribution and shape of the individual soil grains, including the ability or inability of soil grains to aggregate or bond together. Soil behavior is also influenced by the arrangement of soil grains into soil structure, the compactness of the soil structure, and the moisture condition. The mode of deposition, history of loading, and the environment imposed on a soil have a marked effect on these factors.

Since the factors that influence the properties of soils are interrelated, it is possible to determine general soil behavior from the physical make-up of the soil. Soil classification systems have been developed to group soils into categories by determining the particle size distribution of the soil and other easily measured physical features. The groupings within these systems group soils having similar properties, and they are identified with a certain name or symbol.

The fundamental purpose of soil classification schemes is to organize and convey knowledge of the general characteristics of a particular soil in a brief, concise manner. A good soil classification system describes soil in well-understood terms that indicate soil properties related to behavior or performance. The system should be applicable for field determination as well as more precise laboratory measurements. It should employ a simple system of notations for recording soil profiles and conveying information.

There are several types of soil classifications systems. Each has been developed by classifying soils based on their common characteristics relevant to the purpose of the classification. Engineering classification systems permit interpretation of the use of soil as a building material or foundation for structures, dams, or highways. The two main engineering classification systems are the Unified Soil Classification System (Unified)[1] and the American Association of State Highway and Transportation Officials System (AASHTO).[2]

The pedological system of soil classification[3] (Soil Taxonomy) was developed to assist in the mapping and interpretation of soil for urban and rural land use planning. It is the system used in the mapping of soils and preparation of county or area soil survey reports by the Soil Conservation Service of the U.S. Department of Agriculture. The pedological system deals with soil as a natural body and identifies and classifies soils based on properties that result from the interaction of the soil forming factors such as parent material, topography, drainage, climate, biota, and age. Where these soil forming factors are identical, the soils will usually be similar in their properties.

The U.S. Department of Agriculture soil textural classification system is used to identify soil textural classes for the pedological classification. The textural system is based on particle size distribution only and has been related to many uses of soil for agricultural purposes.

PARTICLE SIZE — TEXTURAL TRIANGLE

The U.S. Department of Agriculture texture classification system is contained in U.S. Department of Agriculture Handbook No. 18, Soil Survey Manual.[4] Much of the following procedure and explanation for classifying soil in this system is taken directly from that publication.

Soil texture refers to the relative proportions of the various size groups of individual soil

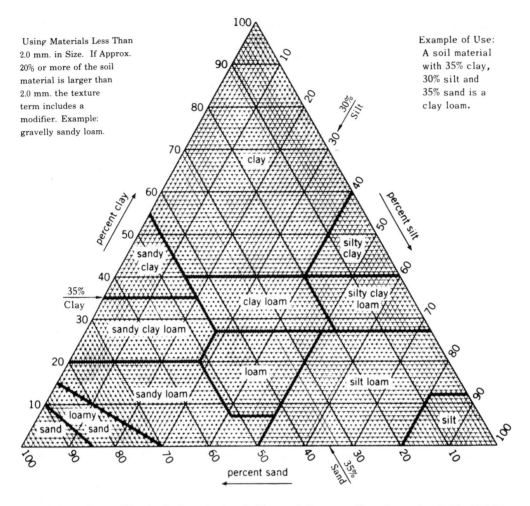

Using Materials Less Than 2.0 mm. in Size. If Approx. 20% or more of the soil material is larger than 2.0 mm. the texture term includes a modifier. Example: gravelly sandy loam.

Example of Use: A soil material with 35% clay, 30% silt and 35% sand is a clay loam.

FIGURE 1. Texture Triangle, for determination of soil textural class name. Example: a soil material with 35% clay, 30% silt, and 35% sand is a clay loam. (These are percent weights of materials less than 2 mm in size.)

    Sand: 25% or more very coarse, coarse, and medium sand (but less than 25% very coarse and coarse sand) and less than 50% fine or very fine sand.

    Fine sand: 50% or more fine sand; or less than 25% very coarse, coarse, and medium sand and less than 50% very fine sand.

    Very fine sand: 50% or more very fine sand.

2.   *Loamy sands*: soil material that contains at the upper limit 85 to 90% sand; percentage of silt, plus one and one half times the percentage of clay is 15 or more; at the lower limit, it contains not less than 70 to 85% sand; percentage of silt, plus twice the percentage of clay, does not exceed 30.

    Loamy coarse sand: 25% or more very coarse and coarse sand and less than 50% any other one grade of sand.

    Loamy sand: 25% or more very coarse, coarse, and medium sand (but less than 25% very coarse and coarse sand), and less than 50% fine or very fine sand.

    Loamy fine sand: 50% or more fine sand; or less than 25% very coarse, coarse, and medium sand and less than 50% very fine sand.

    Loamy very fine sand: 50% or more very fine sand.

3.   *Sandy loams*: soil material that contains 20% clay or less — and the percentage of silt

Table 1
SIZE LIMITS OF U.S.
DEPARTMENT OF
AGRICULTURE SOIL
SEPARATES

Name of separate	Diameter range (mm)
Very coarse sand	2.0—1.0
Coarse sand	1.0—0.5
Medium sand	0.5—0.25
Fine sand	0.25—0.10
Very fine sand	0.10—0.05
Silt	0.05—0.002
Clay	Less than 0.0002

grains in a mass of soil. Specifically, it refers to the percentage, on a weight basis, of sand, silt, and clay in the fraction of the soil that has particles smaller than 2 mm in size.

Soil separates are the individual size groups of particles. The main size groups are *sand, silt,* and *clay.* The sand group is divided into several particle size ranges from very coarse to very fine. Sometimes, larger sizes (rock fragments) are indicated by descriptive terms, but only the groups of particles below 2 mm in diameter are included as soil separates. The scheme used in the U.S. Department of Agriculture is shown in Table 1. Particle size analyses of soils in Soil Survey Laboratories of the U.S. Department of Agriculture are reported in this system.

Rarely, if ever, do soils consist wholly of one separate. Classes of soil texture are based on different combinations of sand, silt, and clay. The basic classes in order of increasing proportions of the fine separates are sand, loamy sand, sandy loam, loam, silt loam, silt, sandy clay loam, clay loam, silty clay loam, sandy clay, silty clay, and clay. Those with the term sand in the name, except for sandy clay loam and sandy clay, are modified for very fine, fine, and coarse depending on the dominance of the specific sand separate.

The basic soil textural class names now used are defined in terms of particle size distribution determined by particle size analysis.[5] The soil texture classes are shown in graphic form in Figure 1. This three-sided graph is often called the texture triangle and is used by plotting the percentages of clay (particles less than 0.002 mm in size), silt (0.002 to 0.05 mm), and sand (0.05 to 2.0 mm). Percentages are calculated by separating a representative sample of the soil into the given size ranges and determining the ratio of the weight in each separate to the total weight of the sample smaller than 2 mm. The sum of the three percentage values should be 100, making it possible to use the chart with two of the three values. Clay percentage values are plotted on the horizontal lines with the scale on the left side of the texture triangle (Figure 1), sand percentage values on diagonal lines sloping from the base upward to the left, and silt percentage values on diagonal lines sloping from the right edge downward to the left. The textural class names are indicated in the various blocks outlined by heavy lines on the graph.

Definitions of the soil textural classes, according to size distribution of mineral particles less than 2 mm in diameter, are as follows:

1.   *Sands*: soil material that contains 85% or more of sand; percentage of silt, plus one and one half times the percentage of clay, does not exceed 15.
     Coarse sand: 25% or more very coarse and coarse sand and less than 50% any other one grade of sand.

plus twice the percentage of clay exceeds 30 — and 52% or more sand; or less than 7% clay, less than 50% silt and between 43 and 52% sand.

Coarse sandy loam: 25% or more very coarse and coarse sand and less than 50% any other one grade of sand.

Sandy loam: 30% or more very coarse, coarse and medium sand (but less than 25% very coarse and coarse sand) and less than 30% very fine or fine sand.

Fine sandy loam: 30% or more fine sand and less than 30% very fine sand; or between 15 and 30% very coarse, coarse, and medium sand.

Very fine sandy loam: 30% or more very fine sand; or more than 40% fine and very fine sand — at least half of which is very fine sand — and less than 15% very coarse, coarse, and medium sand.

4. *Loam*: soil material that contains 7 to 27% clay, 28 to 50% silt, and less than 52% sand.
5. *Silt loam*: soil material that contains 50% or more silt and 12 to 27% clay; or 50 to 80% silt and less than 12% clay.
6. *Silt*: soil material that contains 80% or more silt and less than 12% clay.
7. *Sandy clay loam*: soil material that contains 20 to 35% clay, less than 28% silt, and 45% or more sand.
8. *Clay loam*: soil material that contains 27 to 40% clay and 20 to 45% sand.
9. *Silty clay loam*: soil material that contains 27 to 40% clay and less than 20% sand.
10. *Sandy clay*: soil material that contains 35% or more clay and 45% or more sand.
11. *Silty clay*: soil material that contains 40% or more clay and 40% or more silt.
12. *Clay*: soil material that contains 40% or more clay, less than 45% sand, and less than 40% silt.

In addition to these soil textural class names, modified according to the size group of the sand fraction, other terms are also added as modifiers.

Significant proportions of rock fragments (coarse/very coarse, 2mm) are recognized by an appropriate adjective in the textural soil-class name. Such fragments are regarded as a part of the soil mass. They influence moisture storage, infiltration, and runoff. They also influence root growth, especially through their dilution of the mass of active soil. They protect the fine particles from washing and blowing. They are moved with the soil mass in tillage.

Rock fragments are described in terms that characterize their sizes and shapes. The accepted adjectives to include in textural soil class names and the size limits of classes of rock fragments are set forth in outline form in Table 2. For example, gravelly loam has about 15% or more gravel by volume in the whole soil mass. The basic soil textural class name, however, is determined from the size distribution of the material below 2 mm in diameter, that is, the percentages used for the standard soil class designations are met after the coarse fragments are excluded.

The adjectives listed in the third column of Table 2 are incorporated into the soil textural class designations of horizons when the soil mass contains more than 15% by volume. These adjectives become parts of soil textural names. Another subdivision is made of these fragments at about 35% by volume to give, for example, gravelly loam (15 to 35% gravel), very gravelly loam (35 to 60% gravel), and extremely gravelly loam (more than 60% gravel). Where the fragments make up 90% or more of the soil mass by volume, it is commonly referred to as gravel. The other defined rock fragments are handled similarly.

The adjectives for stones and boulders listed in the third column of Table 2 are incorporated into the soil textural class designation of horizons. The terms ''stony'', ''very stony'', ''bouldery'', and ''very bouldery'' are used to describe the proportion of the volume of a soil horizon they occupy, with no implication that an entire stone or boulder is within the limits of the horizon.

Table 2
TERMS FOR ROCK FRAGMENTS

Shape[a] and size	Noun	Adjective
Rounded, subrounded, angular, or irregular		
0.2—7.6 cm diameter	Gravel[b]	Gravelly
0.2—0.5 cm diameter	Fine gravel	Fine gravelly
0.5—2 cm diameter	Medium gravel	Medium gravelly
2.0—7.6 cm diameter	Coarse gravel	Coarse gravelly
7.6—25 cm diameter	Cobble	Cobbly
25—60 cm diameter	Stone	Stony
>60 cm diameter	Boulder	Bouldery
Flat		
0.2—15 cm long	Channer	Channery
15—38 cm long	Flagstone	Flaggy
38—60 cm long	Stone	Stony
>60 cm long	Boulder	Bouldery

[a] If significant to classification or interpretation, the shape of the fragments is indicated: "angular gravel," "irregular boulders," etc.

[b] A single fragment is called a "pebble."

Rock fragments are estimated in the field on a volume basis. For engineering purposes, it is necessary to convert volume percentage to weight percentage.

Muck, peat, and mucky peat are used in place of the textural class names in organic soils — muck for well-decomposed soil material, peat for raw undercomposed material, and mucky peat for intermediate materials. The word "mucky" is also used as an adjective on the textural class name for mineral soils that contain roughly 10% or more of decomposed organic matter. Designating the texture of a soil as "mucky loam" or "mucky silt loam" implies the presence of enough organic matter that the material has some properties of organic soil combined with properties of the mineral material.

The determination of soil texture is made in the field mainly by feeling the soil with the fingers. This is sometimes supplemented by examination under a hand lens. Field determination requires skill and experience, but good accuracy can be obtained if field classifications are frequently checked against laboratory classifications. The soil must be well-moistened and rubbed vigorously between the fingers for a proper determination of textural class by feel.

REFERENCES

1. **American Society for Testing and Materials,** *Annual Book of ASTM Standards* , Vol. 04.08 ASTM D 2487, American Society for Testing and Materials, Philadelphia, Pa., 1986, 397.
2. **The American Association of State Highway and Transportation Officials,** *Standard Specifications for Transportation Materials and Methods of Sampling and Testing* (Part 1), American Association of State Highway and Transportation Officials, Washington, D.C., 1978, 184.
3. **Soil Survey Staff,** Soil Taxonomy, Soil Conservation Service, Agriculture Handbook No. 436, U.S. Department of Agriculture, Washington, D.C., 1975.
4. **Soil Survey Staff,** Soil Survey Manual, Agriculture Handbook No. 18, U.S. Department of Agriculture, Washington, D.C., 1962, 205.
5. **Kilmer, V. J. and Alexander, L. T.,** Methods of making mechanical analysis of soils, *Soil Sci.*, 68, 15, 1949.

SOIL MECHANICS

J. L. Smith

INTRODUCTION

Soil mechanics is that branch of engineering mechanics that attempts to quantify soil characteristics and predict the response of soil to external or internal disturbances. A disturbance may be natural, for example, an increase in soil density caused by the impact of raindrops, or caused by man, such as the change in soil properties resulting from tillage.

Classic civil engineering soil mechanics provides solutions to problems of foundation design, slope stability, earth retaining structures, and settlement of structures. These specific problems seldom require the degree of attention in agricultural applications that they require in civil engineering applications. However, the same methods can be used to evaluate soil, and concepts and approaches similar to those used in civil engineering can be used in the analytical analysis of some agricultural engineering soil mechanics problems. For example, the same soil or parameters are used to describe soil conditions and strength in both civil and agricultural engineering applications. Further, classic civil engineering retaining wall theory has been used to predict the forces acting on tillage tools.[1]

Agricultural engineers use the term "soil dynamics" to define the specific area of soil mechanics related to the behavior of soils subjected to moving force systems. Tillage and traction are soil dynamics problems.

Most agricultural engineering soil mechanics and soil dynamics problems deal with surface soils and seldom require consideration of soil behavior to a depth of more than 1 M. In this zone, the soil is typically unsaturated, extremely variable and is usually difficult to quantify and/or describe.

The application of soil mechanics to agricultural engineering problems involves three basic steps. First, it is necessary to quantify and/or describe the soil conditions in the area of zone of interest. Water content and density are examples of parameters commonly used to describe soil conditions. The second step is to develop and/or select an appropriate analytical model for predicting the response of the soil to the anticipated disturbance. Most analytical models used in agricultural engineering soil mechanics are of a semiempirical nature.

The final step in the solution of an agricultural engineering soil mechanics problem is to apply and/or interpret the solution to the real field situation. This step is unquestionably the most difficult and requires considerable practical experience. In this regard, it is important to recognize that the accuracy of the analytical solution to a well-developed and defined machine design problem is often no better than 10 to 20% of the experimentally observed result. Therefore, one should expect neither extreme precision nor accuracy in soils where quantification of material parameters is many times more difficult, and the validity of analytical models is always suspect.

SOIL PARAMETERS

Agricultural soils are an assemblage of four different and distinct phases: water, air, organic material, and the soil particles themselves. These phases interact in an extremely complex manner to form the soil structure, and thus soils also exhibit extremely complex and variable behavior.

The composition, size, and distribution of different sized particles in the soil mass, the quantity and distribution of water within the soil structure and the presence of different ions, particularly sodium and calcium, in the water, all have an important effect on soil behavior.

If a force system is applied to an assembly of gravel size particles, water may alter the frictional forces developed as the gravel particles attempt to slide relative to each other. However, in an assembly of clay-size particles, water influences the magnitude and type of chemical bonds between the particles. In addition, the presence of different ions may alter the arrangement of the clay particles and thus the soil behavior. Calcium ions tend to promote a flocculated (honeycomb) arrangement of clay particles, whereas sodium ions tend to promote a dispersed (flat plate) particle arrangement.

Soil parameters can be classified in two groups: index and strength. Index parameters provide a descriptive code that includes the soil type and its condition. Water content, density, and classification are examples of index parameters.

Soil strength parameters include a measure of the maximum capability of the soil to resist loads and its deformation response to loads. Also, adhesive and frictional forces developed between soils and other materials (soil and a steel tillage tool) are treated as soil strength parameters.

Soil Classification

Soil classification and classification systems are discussed in detail in another section of this volume. The discussion here will be limited to the use of soil classification systems in soil mechanics.

The ultimate goal of engineering soil classification is to group soils according to their engineering behavior. That is, soils that exhibit similar responses to similar disturbances should be classified together. Most soil classification schemes are based on the percentage of particles falling within size ranges designated clay, silt, sand, and gravel. In agricultural engineering, both the U.S. Department of Agriculture and Unified classifications are used. Of the two, the Unified is preferred because it is more widely used in the engineering literature.

Classification of fine grained soils in the Unified system is based on Atterberg limits rather than grain size, and the presence of organic materials is noted. Although the Atterberg limits[2] appear arbitrary and awkward, they are based on engineering soils tests, which are widely accepted and precisely reproducible. Hence, classification according to the Unified system enables an engineer familiar with the system to interpret observed soil behavior based on his experience with similarly classified soils.

Soil classification is not changed by soil disturbance. Thus it is not included in the analysis of soil behavior. (An exception might be the addition of organic material to the soil.) However, soil classification is often helpful in selecting an appropriate analytical model and is extremely useful in interpreting the value and applicability of the results of soil mechanics problems to field soils. Soil classification should always be reported in technical publications to aid others in interpreting and correlating observed soil behavior.

Soil Conditions

Soil conditions can be described in terms of water content, density, porosity, voids ratio, degree of saturation, organic matter content, standard compaction test results (e.g., Proctor), and specific gravity. Definitions of most of the above terms are given in Figure 1 for a soil consisting of three phases — air, water, and solids.

Due to the variability of most soils, soil conditions must be determined by direct measurement. Water content and dry density are usually adequate to describe soil conditions in agricultural engineering soil mechanics problems. Water content is measured by determining the loss in water resulting from drying a soil sample for 24 hr at just over 105°C. Dry density is measured by removing and weighing a sample of known volume from the soil mass and correcting the weight for the quantity of water contained in the sample.

Standard compaction tests are typically used for preparation of design specifications for

SOIL PHASES

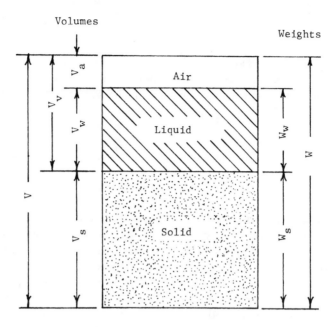

SOIL CONDITIONS

$$\text{Water Content} = \frac{W_w}{W_s}$$

$$\text{Dry Density} = \frac{W_s}{V}$$

$$\text{Bulk Density} = \frac{W}{V}$$

$$\text{Porosity} = \frac{V_v}{V}$$

$$\text{Voids Ratio} = \frac{V_v}{V_s}$$

$$\text{Degree of Saturation} = \frac{V_w}{V_v}$$

$$\text{Specific Gravity} = \frac{W_s/V_s}{\gamma_o}$$

γ_o = specific gravity of water at 4°C

FIGURE 1.   Relationships for soil conditions.

earth structures. The procedure consists of compacting soil having a known water content in a container using a specific procedure.[2] The procedure is repeated using several different soil water contents. The dry density of the soil increases with increasing water content until the maximum dry density, referred to as optimum, is reached. After optimum dry density, the dry density decreases with increasing soil water content. Compacted soil has the highest strength and is thus most stable when the dry density and water content are near optimum.

It should be noted that different standard compaction tests provide somewhat different values for optimum dry density and water content. Therefore, experience is required in selecting the compaction test procedure and for interpreting test results.

Soil Strength

Soil strength parameters include a measure of the maximum capability of the soil to resist loads, its response to dynamic loads, and its deformation response to loads. Because of the similarity of the analysis, soil-metal adhesion and friction will be included in the discussion of soil strength.

Shear Strength

Soil shear strength is normally evaluated using the Coulomb failure criterion:

$$\tau = C + \sigma \tan \phi \tag{1}$$

where τ is the shear stress at failure, C is the apparent cohesion (units of stress), σ is the normal stress on the plane of failure and ϕ is the apparent angle of internal friction. Soil strength parameters are thus C and ϕ. Other failure criteria or parameters, typically more complicated, have not proved useful. This is expected since most soils exhibit anisotropic behavior. The orientations of stress systems and stress distributions applied by strength measurement devices and their relation to stress systems applied by structures or machines is neglected in most agricultural (and civil) engineering soil mechanics problems. This is the same as assuming the soil is isotropic and homogeneous.

Numerous soil strength measuring devices have been proposed. They range in sophistication from the simple grouser plate to the triaxial cell, both shown diagrammatically in Figure 2. All tests have common characteristics in that τ is measured for various values of σ, and C, and ϕ are determined by fitting a straight line to the experimental data.

Bailey and Weber[3] presented conclusive evidence that different shear strength measuring devices produce different soil strength parameters. Geometry of strength measurement devices, the rate of deformation, and the anisotropic behavior of soils are critical to the evaluation of C and ϕ. Selection of a useful soil strength measurement depends almost exclusively upon experience and insights developed into specific problems.

Soil strength parameters determined with a specific device and deformation rate are influenced principally by soil type, water content, and dry density. Therefore, if an evaluation of soil strength is required, a device must be selected and values must be determined experimentally for the given soil and soil conditions. Bailey and Weber[3] give a complete analysis of the commonly used soil strength measurement devices.

Other soil strength parameters have proved useful for developing equations to describe the behavior of soil-machine systems. An excellent example is in the area of oscillating tillage tools. In this application, soil parameters representing perfectly elastic and perfectly plastic components were useful.[4]

Strength-Deformation Characteristics

The simplest criterion to evaluate the shear strength deformation characteristics of agricultural soils is the Coulomb-Micklethwait equation:

$$\tau = (C + \sigma \tan \phi)(1 - e^{-j/K}) \tag{2}$$

where τ, C, and ϕ are as defined for Equation 2, j is the displacement, and K is determined from experimental data.

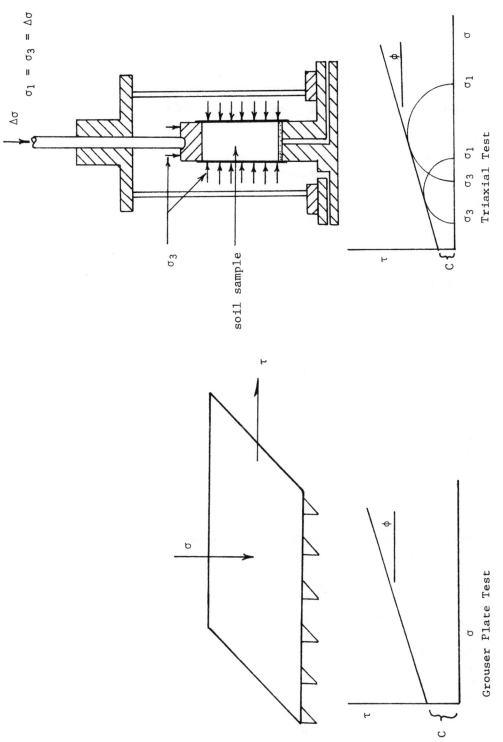

FIGURE 2.   Soil strength tests.

Soil-Metal Adhesion and Friction

Payne and Fountaine[5] proposed an equation similar in form to the Coulomb equation to evaluate the behavior of soil moving over the surface of another material:

$$S_s = C_a + \sigma \tan \delta \tag{3}$$

where S_s is the sliding stress, C_a is the adhesion (units of stress), σ is the normal stress on the sliding surface, and δ is the angle of soil-material friction. As with the Coulomb equation, tests must be conducted to evaluate C_a and δ from measured values of S_s and σ. C_a and δ are thus the soil-material adhesion and friction parameters.

Water content, soil type, and surface roughness affect the parameters C_a and δ. In addition, the value of δ tends to decrease once the soil begins to slide, and it may be necessary to evaluate both dynamic and static friction coefficients.

Soil Cone Penetration Test

The cone penetration test (ASAE-S 3/3.1, ASAE[6]) is an interesting combination of an index and strength parameter test. Its primary advantage is simplicity and speed with which tests can be conducted. Results of the tests are usually presented in terms of the cone index or CI, the force per unit base area required to push the standard cone into the soil at a specified rate. The cone penetration test is very useful in traction studies. This usefulness has resulted from extensive testing and experience, and standardization of the test.

The cone index typically varies with depth. For this reason, experience is required to select the most suitable CI value for a given soil problem. In traction studies, the value most often used is the average CI over a soil depth of 15 cm.

The full potential of the cone penetration test has yet to be developed. It is useful in predicting the hydraulic properties of soils.[7,8] With proper calibration, the cone penetrometer can be used to predict either soil density or water content in a given soil, if either is known, along with penetration resistance.[9]

In the author's experience, the cone penetrometer is sensitive to the shape and size of the cone and the rate of penetration. In view of the simplicity of the test and the need to establish soil parameters useful in tillage studies, where both geometric and rate effects are important, additional research should be conducted with the cone penetrometer.

ANALYTICAL ANALYSIS OF SOIL BEHAVIOR

Analytical solutions are available for many civil engineering soil mechanics problems.[10] Generally, these solutions involve analysis of the static equilibrium of a soil mass.

The most comprehensive analytical analysis of an agricultural engineering soil mechanics problem was published by Bekker.[11] His analysis provided a method for estimating the net tractive effort or thrust that could be developed by a wheel or track. Solutions have also been proposed for many types of tillage tools.[12] Numerous other approaches are discussed in the literature.

Dimensional analysis and/or similitude has been used extensively in soil dynamics studies. Freitag et al.[13] presented an excellent discussion of the application and limitations of these methods.

Prediction of the thrust produced by a wheel has been successfully solved by direct analysis of the behavior of a wheel and by dimensional analysis. The approaches used both evolved from the same governing equation proposed by Bekker:[11]

$$P = f(1 - e^{-s}) \tag{4}$$

where P is the tractive effort or thrust and S is the wheel slip. These developments will be discussed here to illustrate the use of soil parameters in soil mechanics and to demonstrate the selection and application of a specific method of solution in a practical problem.

Wismer and Luth[14] used the form of Equation 4 and dimensional analysis to develop the following equation:

$$P = W[0.75(1 - e^{-0.3C_n S})] \qquad (5)$$

where P is the gross wheel pull, W is the dynamic vertical load on the wheel, and C_n = CI bd/W, where CI is the soil cone index, b is the wheel width, and d is the wheel diameter. The net tractive effort can be obtained by subtracting the rolling resistance from the gross wheel pull. The rolling resistance R is estimated using the empirical equation:

$$R = \frac{1.2}{C_n} + 0.04 \qquad (6)$$

Equations 5 and 6 are simplified from equations previously developed at Waterways Experiment Station.[15] Through experience and insight, Wismer and Luth selected important independent variables and combined them into dimensionless terms that were then adapted to the original form of the governing equation (Equation 4). The resulting equation was then corrected empirically (Equation 6) for rolling resistance. Note that only cone penetration test results are required in Equations 5 and 6 to describe soil conditions.

Bekker's[11] solution to the same problem was developed by multiplying the area of contact of the tire by the shear strength of the soil (Equation 1):

$$P = 0.78 \, b\ell \left[C + \frac{W}{0.78b\ell} \tan \phi \right] \qquad (7)$$

where b and ℓ are the major and minor axles, respectively, of the elliptical contact area between the tire and soil. The complete equation proposed by Bekker[11] involved slip as a variable.

Note that Equation 7 is for the maximum gross wheel pull, i.e., large values of slip. Also, C and ϕ must be determined from a soil shear strength test. Bekker[11] estimated the rolling resistance by equating the work done in deforming the soil to the energy required to overcome rolling resistance. One form of the resulting equation is

$$R = \frac{2}{(n + 1)(K_c + bK_\phi)^{1/n}} \left(\frac{W}{2\ell}\right)^{\frac{n + 1}{n}} \qquad (8)$$

where K_c, K_ϕ, and n are soil parameters determined by pushing two different sized flat plates into the soil. Bekker[11] also provided equations to correct for other factors in the system.

In the author's experience, the Wismer and Luth[14] approach (Equations 5 and 6) and the Bekker[11] approach (Equations 7 and 8) provide approximately the same estimate of the net tractive force available from a wheel in a single soil type and condition. Experimental results compare with the estimates to within ±20% in most applications.

APPLICATION OF SOIL MECHANICS

The validity and application of soil mechanics problems is as varied and complex as the problems themselves. Experimental system evaluation is invaluable in applying the results of analytical solutions to the field. In fact, as illustrated in the preceding section, experimental measurements may also be invaluable in developing the analysis.

Selection of a particular method of analysis, once it has proved useful, may depend upon many factors. The vast capability and capacity of modern instrumentation and data collection and analysis systems should never be overlooked.

Referring again to the previous comparison of the Wismer and Luth[14] and Bekker[11] equations, it was stated that both solutions provide approximately the same results. At first glance, one might prefer the Bekker solution since it is more rigorous mathematically (although some of the assumptions involved may be questionable). However, in a practical situation, the Wismer and Luth approach is far superior, simply because it requires only a single and very simple test to characterize the soil.

The Bekker[11] approach requires five soil parameters that must be determined experimentally using somewhat complicated and cumbersome equipment. This limits the number of tests that can be made in a given area and/or increases the cost of obtaining the information. On the other hand, the cone penetrometer used in the Wismer and Luth[14] approach is a very simple and quick test. In the same time interval, more cone index values can be obtained on a given land area than shear strength parameters. Also, the equipment is less costly, requires less skilled operators, and is less likely to experience failures associated with experimental equipment. The cone penetrometer can thus be used to obtain a more representative evaluation of the soils in a given area. In the respect, the Wismer and Luth[14] approach should provide a more representative and less costly estimate of the tractive performance of a wheel in the given area.

The above example is one of many that could be used to illustrate the selection and application of an approach to an agricultural engineering soil mechanics problem. It serves to illustrate the fact that there is no substitute for thoughtful consideration of the details involved in obtaining useful experimental information. The value of a solution is often improved by taking more numerous, less sophisticated measurements compared to a few precise measurements. Further, there is a continued need to assess and improve upon the analytical analysis of soil mechanics problems through correlation with field experiences.

REFERENCES

1. **Siemens, J. C., Weber, J. A., and Thornburn, T. H.,** Mechanics of soil as influenced by model tillage tools, *Trans. ASAE,* 8(1), 1, 1965.
2. *ASTM Standards, Part 19,* American Society of Testing and Materials, Philadelphia, Pa., updated periodically.
3. **Bailey, A. C. and Weber, J. A.,** Comparison of methods of measuring soil shear strength using artificial soils, *Trans. ASAE,* 8(2), 153, 1965.
4. **Yow, J. and Smith, J. L.,** Sinusoidal vibratory tillage, *J. Terramech.,* 13(4), 211, 1976.
5. **Payne, P. C. J. and Fountaine, E. R.,** The mechanism of scouring for cultivation implements, *Natl. Inst. Agric. Eng. Tech. Memo.,* 116, 1, 1954.
6. **ASAE, Soil cone penetrometer, ASAE Standard: ASAE S-313.1,** *Agricultural Engineers Yearbook,* American Society of Agricultural Engineers, St. Joseph, Mich., 1979.
7. **Chan, D. T. and Hendrick, J. G.,** Using a Cone Penetrometer to Predict Soil Density and Hydraulic Conductivity, paper presented at Fall National Meeting, American Geophysical Union, San Francisco, December 2 to 5, 1968.
8. **Khalid, M. and Smith, J. L.,** Control of furrow infiltration by compaction, *Trans. ASAE,* 20(6), 650, 1978.
9. **Smith, J. L.,** The Effect of Mode of Vibration and Blade Angle on the Performance of a Simple Vibratory Tillage Tool, Ph.D. dissertation, University of Minnesota, Minneapolis, 1971.
10. **Lambe, T. W. and Whitman, R.,** *Soil Mechanics, SI Version,* John Wiley & Sons, New York, 1979.
11. **Bekker, M. G.,** *The Mechanics of Vehicle Mobility,* University of Michigan Press, Ann Arbor, 1957.

12. **Gill, W. R. and Vandenberg, G. E.,** Soil dynamics in tillage and traction, *Agriculture Handbook No. 316,* Agricultural Research Service, U.S. Department of Agriculture, Washington, D.C., 1967.
13. **Freitag, D., Schafer, R. L., and Wismer, R. D.,** Similitude studies of soil machine systems, *J. Terramech.,* 7(2), 25, 1970.
14. **Wismer, R. D. and Luth, H. J.,** Off road traction prediction for wheeled vehicles, *Trans. ASAE,* 17(1), 8, 1974.
15. **Rula, A. A. and Nuttall, C. J.,** An Analysis of Ground Mobility Models, Tech. Rep. M-72-4, U.S. Army Engineers Waterways Experiment Station, Vicksburg, Miss., 1971.

DENSITIES

W. W. Brixius

Soil density is defined as the weight of a unit volume of soil expressed on either a wet basis (including soil and water) or on a dry basis (soil only, most common). Bulk density is used most often in agriculture when referring to soil compaction or, less commonly, to soil strength. Density is dependent on the soil texture, moisture content, and previous soil history with respect to compactive efforts.

The value of density relates directly to the relative amount of soil, air, and water in the soil mass. Since the specific gravity of the solid particles (clay, silt, sand) ranges between 2.60 and 2.80, the amount of air voids and water can be calculated, knowing both the soil density and moisture content of the soil. The relative amounts of air and water in turn have been related to plant growth, root penetration, and crop yield in a number of research programs.

UNDISTURBED *(IN SITU)* SOIL

Undisturbed or *in situ* soils refer to soils in their natural or original state as opposed to soils having been recently tilled, remolded, or distorted in some degree. In undisturbed soils, soil dry density relates strongly with the distribution of particle size. A well-graded soil, that is, one with a mix of small, medium, and large particle size, will normally have a relatively high soil density. Voids between the larger particles (sand) are filled with small particles (silt, clay) resulting in a high weight per unit volume. Typical density values for a variety of soil textures in wet-season conditions[5] are contained in Table 1. Dry density ranges from 1.01 to 1.64 (63 to 102 lb/ft³) and includes data from over 700 sites throughout the U.S.

This is further shown in Figure 1[5] using the U.S. Department of Agriculture Textural Classification Chart. Maximum densities were obtained from predominantly sandy soil with the maximum value containing approximately 25% clay. Densities decreased very rapidly with an increase in clay content above 25%.

In the same study, it was found that mean density of a particular soil type was slightly higher in low topography (water table within 4 ft of surface) than in high topography. It was felt the soil structure was altered in the lower areas due to settlement in compaction of soil because of fluctuation of the water table level.

DISTURBED SOIL

Soil density may be artificially changed through tillage, traffic, or other remolding means. In construction, minimum dry densities are specified for compacted soils for road beds, foundations, etc. The degree of soil density is most strongly affected by soil texture, moisture content, and compactive effort.

Several laboratory compaction methods have been developed to determine moisture-density relationships. The most common method is the Proctor test where soil is compacted into a mold with a volume of 1/30 ft³ in three layers, by 25 blows from a 5.5-lb hammer falling 12 in. Figure 2 shows how soil dry density varies with moisture content for two different soils.

Generally, at low water contents, the degree of compaction increases with an increasing amount of water. The added moisture helps to soften and plasticize the soil, which aids in the densification process. Beyond an optimum point, however, dry density drops off rapidly

Table 1
TYPICAL DRY DENSITIES FOR
UNDISTURBED SOIL TYPES

USCS type	Dry density (g/cc)	USDA type	Dry density (g/cc)
SM-SC	1.62	SCL	1.64
SC	1.60	S	1.54
SP-SM	1.57	LS	1.50
SM	1.50	SL	1.49
CL-ML	1.50	CL	1.48
CL	1.46	SiCL	1.46
ML	1.37	L	1.45
CH	1.36	SiC	1.42
OL	1.32	SiL	1.39
MH	1.12	Si	1.37
OH	1.01	C	1.32
		SC	1.27

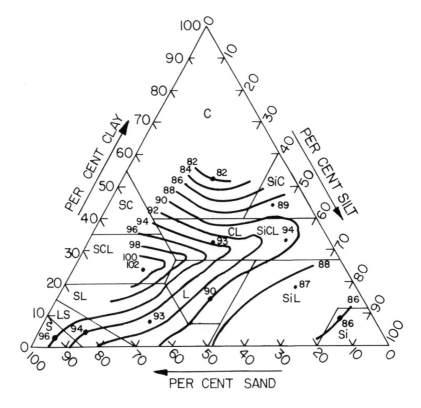

FIGURE 1.   Dry density in pounds per cubic foot under wet-season conditions.

with the addition of water. At this stage, the soil is near saturation (all voids filled with water) and there is little air void left to compact.

Optimum moisture content, as well as maximum soil density, varies with soil type. A silty clay soil, for example, holds more water than sand, meaning optimum moisture content is high (Figure 2). Also, highest densities are obtained with a well-graded soil where smaller particles surround larger ones in filling voids.

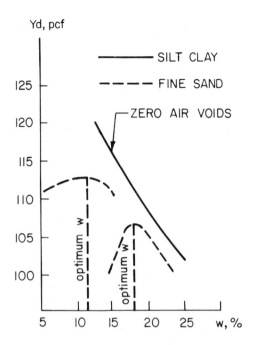

FIGURE 2. Moisture-density curves for two soils. (From Wu, T. H., *Soil Mechanics,* 2nd ed., Allyn and Bacon, Rockleigh, N. J., 1967. With permission.)

Soil densification or compaction under agricultural conditions thus is dependent on soil type, moisture content, and wheel/implement load. During wet seasons, where moisture contents may be near optimum, compaction may be a considerable problem. For this reason, traffic is discouraged when moisture content is too high. Consequences of soil compaction regarding plant growth and crop yields are discussed in the section "Compaction-Growth Effects".

MEASUREMENT OF SOIL DENSITY

Soil density is normally measured in one of two methods.[4] The first method is to determine the weight and volume of a sample of the soil. The second is to relate the soil's conductivity for the passage of fluids or for nuclear radiation. The methods are described in detail in a number of references and are described briefly here.

The most common weight-volume method is to weigh core samples of known volume and calculate density. Core samples are usually designed to work effectively in a particular soil condition. The main problem is in acquiring a sample with as little disturbance to the sample as possible.

In cases where a core sample cannot be obtained due to the sampling process affecting the volume, the volume of the hole evacuated in the soil must be measured, using some volume displacement technique. These techniques include a sand density, oil displacement, and water balloon methods. A complete description of the methods, drawbacks, and advantages can be obtained from the references.

Radiation methods have been improved in recent years in achieving accurate analogues to soil density. Basically, gamma rays are emitted from a radioactive isotope into the soil. These rays interact with the soil mass so that the gamma ray energy remaining after passage through the soil relates to the mass density. Numerous papers describe specifications of various devices. Technology and experience have greatly improved the accuracy of these methods in recent years. Density measurements are far quicker than with the weight-volume methods; however, equipment is much more costly.

REFERENCES

1. **Tschebotarioff, G. P.,** *Soil Mechanics, Foundations and Earth Structures,* 6th ed., McGraw-Hill, New York, 1956.
2. **Gill, W. R. and Vanden Berg, G. E.,** Soil Dynamics in Tillage and Traction, Agricultural Handbook No. 316, Agricultural Research Service, U.S. Department of Agriculture, Washington, D.C., 1967.
3. **Harris, W. L.,** The soil compaction process, in *Compaction of Agricultural Soils,* ASAE Monograph, Barnes, K. K., Carleton, W. M., Taylor, H. M., Throckmortor, R. I., and Vanden Berg, G. E., Eds., American Society of Agricultural Engineers, St. Joseph, Mich., 1971, chap. 1.
4. **Freitag, D. R.,** Methods of measuring soil compaction, in *Compaction of Agricultural Soils,* ASAE Monograph, Barnes, K. K., Carleton, W. M., Taylor, H. M., Throckmortor, R. I., and Vanden Berg, G. E., Eds., American Society of Agricultural Engineers, St. Joseph, Mich., 1971, chap. 3.
5. **Meyer, M. P. and Knight, S. J.,** Soil Classification, Tech. Memo. No. 3-240, 16th Suppl. U.S. Army Engineers Waterways Experiment Station, Vicksburg, Miss., 1961.
6. **Wu, T. H.,** *Soil Mechanics,* 2nd ed., Allyn and Bacon, Rockleigh, N.J., 1967.
7. PCA Soil Primer, Portland Cement Association, Chicago, 1962.
8. **Chancellor, W. J.,** Compaction of Soil by Agricultural Equipment, Bull. 1881, Division of Agricultural Sciences, University of California, Berkeley, 1977.
9. **Raghavan, G. S. V., McKyes, E., Stemshorn, E., Gray, A., and Beaulieu, B.,** Vehicle compaction patterns in clay soil, *Trans. ASAE,* 20(2), 1977.

Agricultural Machinery — Functional Elements

TERMINOLOGY AND DEFINITIONS FOR AGRICULTURAL TILLAGE IMPLEMENTS AND AGRICULTURAL EQUIPMENT*

TERMINOLOGY FOR AGRICULTURAL EQUIPMENT

1.1. Agricultural field equipment: Agricultural tractors, self-propelled machines, implements, and combinations thereof designed primarily for agricultural field operations.

1.1.1. Agricultural tractor: A self-propelled wheeled or track-laying machine, designed primarily to provide the tractive power to pull, push, propel, carry and/or provide power to implements designed primarily for agricultural usage. Agricultural tractors have 15 kW (20 hp) or more net engine power per Society of Automotive Engineers Standard J816, Engine Test Code-Spark Ignition and Diesel.

1.1.2. Agricultural implement: An implement that is designed to perform agricultural operations.

1.1.2.1. Towed implement: An implement that is pulled by a tractor and is usually equipped with wheels for transport.

1.1.2.2. Mounted implement: An implement which is mounted directly on the tractor and is carried by the tractor during transport.

1.1.2.3. Semi-mounted implement: An implement which is partially mounted on the tractor and partially carried on wheels during operation and/or transport.

1.1.3. Self-propelled machine: An implement designed with integral power unit to provide both mobility and power for performing agricultural operations. Terminology for some self-propelled machines follows.

1.1.3.1. Self-propelled beet harvester: A self-propelled machine which digs and conveys sugar beets to an attached bin or into an accompanying truck or wagon.

1.1.3.2. Self-propelled combine: A self-propelled machine for harvesting a wide variety of grain crops. Normally the machine gathers the crop.

1.1.3.3. Self-propelled cotton picker: A self-propelled machine for collecting cotton from open bolls on the stalk usually consisting of picking heads equipped with revolving spindles or other picking means, a conveying means, and a bin for carrying the picked cotton.

1.1.3.4. Self-propelled forage harvester: A self-propelled machine which gathers and chops forage crops. The machine usually has a blower type discharge which loads the chopped material into accompanying wagons and trucks.

1.1.3.5. Self-propelled hay baler: A self-propelled machine which picks up and compresses loose hay into compact bales and secures them with wire or twine. Bales are discharged to the ground or to an accompanying wagon or truck.

1.1.3.6. Self-propelled high clearance sprayer: A self-propelled machine which carries a storage tank, pump and spray heads for spraying crops which require high clearance. Crop clearance of the machine is usually over 1220 mm (48 in.).

1.1.3.7. Self-propelled windrower: A self-propelled machine which cuts and gathers standing hay or grain into uniform rows for drying and pickup. In haying operations, the machine normally includes a conditioning attachment designed to crimp the hay and thus decrease the drying time required.

1.2. Farmstead equipment: Feed handling equipment, manure handling equipment and accessories for such equipment, designed primarily for use in agricultural materials handling operations.

---

* Adapted from ASAE Standard S390.1 and ASAE Standard S414, ASAE Standards 1985.

TILLAGE IMPLEMENT CATEGORIES

2.1. Primary tillage: Tillage which displaces and shatters soil to reduce soil strength, and to bury or mix plant materials and fertilizers in the tilled layer. Primary tillage is more aggressive, deeper, and leaves a rougher surface relative to secondary tillage.

2.1.1. Examples of primary tillage implements:

Plows
  Moldboard
  Chisel
  Combination chisel with cutting blades
  Wide-sweep
  Disk
  Bedder
Moldboard listers
Disk bedders
Subsoilers
Disk harrows
  Offset disk
  Heavy tandem disk
Power rotary tillers

2.2. Secondary tillage: Tillage that works the soil to a shallower depth than primary tillage, provides additional pulverization, levels and firms the soil, closes airpockets, and kills weeds. Seedbed preparations are the final secondary tillage operations.

2.2.1. Examples of secondary tillage implements:

Harrows
  Disk
  Spring, spike, coil, or tine tooth
  Knife
  Roller
  Powered oscillatory spike tooth
  Packer
  Ridger
  Leveler
  Rotary ground-driven
Cultivators
  Field or field conditioner
Rod weeders
Roller harrows
Powered rotary tillers
Bed shapers
Rotary hoes

2.3. Cultivating tillage: Shallow post-planting tillage whose principle purpose is to aid the crop by either loosening the soil and/or by mechanical eradication or undesired vegetation.

2.3.1. Examples of cultivating implements:

Row crop cultivators
  Rotary ground-driven
  Spring tooth
  Shank tooth
Rotary hoes
Rotary tillers-strip type, power driven

IMPLEMENT HITCH CLASSIFICATIONS

3.1. Pull
    3.1.1. Wheel mounted
    3.1.2. Drag
    3.1.3. Squadron
3.2. Semi-mounted *(semi-integral)*
3.3. Rear mounted *(three-point integral)*
3.4. Front mounted
3.5. Center mounted
3.6. All tillage tools are not produced in all classifications.

IMPLEMENT FRAME CONFIGURATIONS

4.1. Rigid
4.2. Rigid with rigid wings
4.3. Single folding wing
4.4. Dual folding wings
4.5. Multiple folding wings
4.6. Hinged
4.7. Sectional
4.8. Endways transported
4.9. Wing styles may have mechanical, hydraulic, or no folding assistance.

DEFINITIONS

5.1. Disk Harrow: A primary or secondary tillage implement consisting of two or four gangs of concave disks. Adjustment of gang angle controls cutting aggressiveness. Disk harrow hitches are either rear mounted or pull type.

5.2. Offset Disk Harrow: A primary or secondary tillage implement consisting of two gangs of concave disks in tandem. The gangs cut and throw soil in opposite directions.

5.3. One-Way Disk Harrow: A tillage implement equipped with one gang of concave disks. When mounted in short flexible gang units, the harrow conforms to uneven soil surfaces.

5.4. Moldboard Plow: A primary tillage implement which cuts, partially or completely inverts a layer of soil to bury surface materials, and pulverizes the soil. The part of the plow that cuts the soil is called the bottom or base. The moldboard is the curved plate above the bottom which receives the slice of soil and inverts it. Moldboard plows are equipped with one or more bottoms of various cutting widths. Bottoms are commonly right-hand that turn all slices to the right. Two-way moldboard plows are equipped with right-hand bottoms that are alternately used to turn all slices in the same direction as the plow is operated back and forth across the field.

5.5. Chisel Plow: A primary or secondary tillage implement which shatters the soil without complete burial or mixing of surface materials. Multiple rows of staggered curved shanks are mounted either rigidly, with spring-cushions, or with spring resets. Interchangable sweep, chisel, spike, or shovel tools are attached to each shank. Working width is increased by adding folding wings to the main unit. Combination implements consist of chisel plows with gangs of flat or concave disks or individual rolling coulters preceding the shanks to cut surface residue and vegetation. Chisel plows differ from cultivators by being constructed stronger with wider spaced shanks for primary tillage. (See Field cultivators.)

5.6. Disk Plow: A primary tillage implement with individually mounted concave disk blades which cut, partially or completely invert a layer of soil to bury surface material, and pulverize the soil. Blades are attached to the frame in a tilted position relative to the frame and to the direction of travel for proper penetration and soil displacement. Penetration is increased by the addition of ballast weight. Disk plows are equipped with one or more blades of diameter corresponding to intended working depth. Disk plows are commonly right-hand, but two-way plows are equipped with right-hand and left-hand blades.

5.7. Subsoiler: A primary tillage implement for intermittent tillage at depths sufficient to shatter compacted subsurface layers. Subsoilers are equipped with widely spaced shanks either in-line or staggered on a V-shaped frame. Subsoiling is commonly conducted with the shank paths corresponding to subsequent crops rows. Strong frame and shanks are required for deep operation.

5.8. Bedder-Ridger: A primary tillage implement or a secondary tillage implement for seedbed forming. Bedder tools are either moldboard lister bottoms which simultaneously throw soil in both right-hand and left-hand directions or short disk gangs with two or more disks of equal or varying diameters. Each disk gang throws soil in one direction and is followed by another disk gang throwing soil in the opposite direction to form a furrow. Planting attachments are sometimes added behind a bedder for planting either on top of the beds or in the furrows.

5.9. Field Cultivator: A secondary tillage implement for seedbed preparation, weed eradication, or fallow cultivation subsequent to some form of primary tillage. Field cultivators are equipped with spring steel shanks or teeth which have an integral forged point or mounting holes for replaceable shovel or sweep tools. Teeth are generally spaced 15 to 23 cm (6 to 9 in.) in a staggered pattern. Frame sections are folded upwards or backwards for transport.

5.10. Row Crop Cultivator: A secondary tillage implement for tilling between crops rows. The frame and cultivating tools are designed to adequately pass through standing crop rows without crop damage. Gangs of shanks are often independently suspended on parallel linkages with depth-controlling wheels to provide flotation with the soil surface. Tool options are shanks with shovels or sweeps, spring teeth, and ground-driven rotary finger wheels.

5.11. Harrows: Tillage implements used for seedbed preparation and in some cases light surface cultivation after the seed is planted and before or after the crop emerges. Harrows level the soil surface, enhance moisture retention, pulverize surface clods, and disturb the germination of small weeds. Harrows have staggered teeth of either rigid spikes, coil-spring round wires, flat-spring bars, or S-shaped spring bars.

5.12. Rotary Hoe: A secondary tillage implement for dislodging small weeds and grasses and for breaking soil crust. Rotary hoes are used for fast, shallow cultivation before or soon after crop plants emerge. Rigid curved teeth mounted on wheels roll over the soil, penetrating almost straight down and lifting soil as they rotate. Hoe wheels may be mounted in multiple gangs or as short gangs on spring loaded arms suspended from the main frame.

5.13. Seedbed Conditioner: A combination secondary tillage implement for final seedbed preparation. Typical purpose is to smooth and firm the soil surface for flat-planting.

5.14. Roller Harrow: A secondary tillage implement for seedbed preparation which crushes soil clods and smooths and firms the soil surface. It consists of an in-line gang of ridged rollers, followed by one or more rows of staggered spring cultivator teeth, followed by a second in-line gang of ridged rollers.

5.15. Packer: A secondary tillage implement for crushing soil clods and compacting the soil. Packers consist of one or two in-line gangs of rollers. Roller sections may be lugged wheels or any one of various shaped ridged wheels.

5.16. Rotary Tiller: A primary or secondary tillage implement used for broadcast or strip tillage. Rotary tillers are also used as chemical incorporators prior to planting and as row crop cultivators. They consist of a power-driven shaft, transverse to the direction of travel, equipped with curved knives that slice through the soil, chop surface residue, and mix all materials in the disturbed layer.

TILLAGE TOOL POWER REQUIREMENTS FOR DRAFT TOOLS AND PTO-DRIVEN TOOLS

James G. Hendrick

DRAFT TOOLS*

Draft data as reported herein refer to the force required in the horizontal direction of travel. Only functional draft (soil and crop resistance) is reported. Rolling resistance of transport wheels may have to be added to secure total implement draft.

Moldboard Plows

These plows show a draft per unit cross section of furrow slice for bottoms equipped with high-speed moldboards, coulters, and landsides. Draft is shown in N/cm^2 ($lb/in.^2$), speed (S) in kilometers per hour — km/hr (mi/hr). Typical operating speed is 5.6 to 9.7 km/hr (3.5 to 6 mi/hr).**

Silty clay (South Texas)	$7 + 0.049 \ S^2$
	$(10.24 + 0.185 \ S^2)$
Decatur clay loam	$6 + 0.053 \ S^2$
	$(8.77 + 0.2 \ S^2)$
Silty clay (N. Illinois)	$4.8 + 0.024 \ S^2$
	$(7 + 0.09 \ S^2)$
Davidson loam	$3 + 0.020 \ S^2$
	$(4.5 + 0.08 \ S^2)$
Sandy silt	$3 + 0.032 \ S^2$
	$(4.4 + 0.21 \ S^2)$
Sandy loam	$2.8 + 0.013 \ S^2$
	$(4 + 0.05 \ S^2)$
Sand	$2 + 0.013 \ S^2$
	$(3 + 0.05 \ S^2)$

Multiply values obtained by 1.07 for an added jointer or coverboard. An increase of one soil moisture percentage point at moistures below field capacity can decrease draft 10%. An increase of 0.1 in apparent specific gravity can increase draft 10%. A 0.13 coefficient of variation is common in plow unit draft measurements.

Disk Plows

This plow shows a draft per unit cross section of furrow slice for 66 cm (26 in.) diameter disk, 0.38 rad (22°C) tilt, 0.785 rad (45°C) angle. Draft in N/cm^2 ($lb/in.^2$), speed (S) kilometers per hour — km/hr (mi/hr). Typical operating speed, 5.5 to 9.5 km/hr (3.5 to 6 mi/hr).

Decatur clay	$5.2 + 0.039 \ S^2$
	$(7.6 + 0.15 \ S^2)$

---

* ASAE Standards, Engineering Practices, and Data are informational and advisory only. Their use by anyone engaged in industry or trade is entirely voluntary. There is no agreement to adhere to any ASAE Standard, ASAE Engineering Practice, or ASAE Data. The ASAE assumes no responsibility for results attributable to the application of these Standards, Engineering Practices, and Data. Conformity does not ensure compliance with applicable ordinances, laws and regulations. Prospective users are responsible for protecting themselves against liability for infringement of patents.

** Items in text and in tables not in parentheses are metric units. Items in parentheses are English units.

Davidson loam	$2.4 + 0.045\ S^2$
	$(3.4 + 0.17\ S^2)$

Listers

Draft is per 36 cm (14 in.) bottom at 6.76 km/hr (4.2 mi/hr). Draft shown in Newton per bottom — N/bottom (lb/bottom), depth (d) in centimeters — cm (in.). Typical operating speed is 5.5 to 8 km/hr (3.5 to 5 mi/hr).

Silty clay loam	$21.5\ d^2$
	$(31.2\ d^2)$

Disk Harrows

Draft is per mass (weight) at any speed, typical working depth. Draft is in Newton — N (lb), mass (M) in kilograms — kg (lb). Typical operating speed is 5 to 9.5 km/hr (3 to 6 mi/hr).

Clay	14.7 M
	(1.5 M)
Silt loam	11.7 M
	(1.2 M)
Sandy loams	7.8 M
	(0.8 M)

Typical weights of disk harrows, per unit width,[3] are

Tandem:
    Mounted, 41 to 51 cm (16 to 20 in) diameter, 160 to 210 kg/m (110 to 140 lb/ft)
    Wheel-type, 41 to 66 cm (16 to 26 in) diameter, 240 to 510 kg/m (160 to 340 lb/ft)
Offset, pull type:
    With wheels, 56 to 81 cm (22 to 32 in) diameter, 390 to 890 kg/m (260 to 600 lb/ft)
    No wheels, 61 to 81 cm (24 to 32 in.) diameter, 390 to 890 kg/m (260 to 600 lb/ft)

Chisel Plows and Field Cultivators

Draft in firm soil per tool is spaced at 30 cm (1 ft) includes wheel rolling resistance. Depth (d) = 8.26 cm (3.25 in.). Draft (D) is Newton — N (lb) per tool, speed (S) in kilometers per hour — km/hr (mi/hr). Typical operating speed is 5.5 to 10.5 km/hr (3.5 to 6.5 mi/hr).

Loam (Saskatchewan)	$520 + 49.2\ S$
	$(117 + 17\ S)$
Clay loam (Saskatchewan)	$480 + 48.1\ S$
	$(108 + 16\ S)$
Clay (Saskatchewan)	$527 + 36.1\ S$
	$(118 + 12\ S)$

Draft at depth x (d_x) follows the relationship:

$$D_x = D_{8.26\ cm}\left(\frac{d_x}{8.26}\right)^2$$

$$\left[D_{3.25\ in}\left(\frac{d_x}{3.25}\right)^2\right]$$

Variations in draft of 10% about the mean are common.

One-Way Disk Plow With Seeder Attachment

Draft per unit width, 7.5 cm (3 in.) depth of tillage, includes wheel rolling resistance. Draft is in kilo Newton per meter — kN/m (lb/ft), speed (S) in kilometers per hour — km/hr (mi/hr).

Loam (Saskatchewan)	1.6 + 0.13 S
	(110 + 14 S)
Clay loam (Saskatchewan)	1.7 + 0.13 S
	(120 + 14 S)
Clay (Saskatchewan)	2 + 0.17 S
	(140 + 18 S)

Variations in draft of 10% about the mean are common. Increase draft 3.5% in loam, 7% in clay loam, and 20% in clay for each centimeter increase in depth.

Subsoiler

Draft is per shank per unit depth. Draft is in Newton per shank — N/shank (lb/shank), depth (d) in centimeters — cm (in). Typical operating speed, 5 to 8 km/hr (3 to 5 mi/hr).

Sandy loam	120—190 d
	(70—110 d)
Medium or clay loam	175—280 d
	(100—160 d)

Minor Tillage Tools

Draft is per unit width, average for all soils. Draft is in Newton per meter — N/m (lb/ft), speed in kilometers per hour km/hr (mi/hr).

	Draft	Speed
Land plane	4,400—11,600	5—8
	(300—800)	(3—5)
Spike tooth harrow	440—730	4.8—11.3
	(30—50)	(3—7)
Spring tooth harrow	1,460—2,190	4.8—11.3
	(100—150)	(3—7)
Rod weeder	880—1,830	6.4—9.7
	(60—125)	(4—6)
Roller or packer	440—880	4.8—12
	(30—60)	(3—7.5)

Row Planters

Draft is per row and includes wheel rolling resistance. Draft is in Newton per row — N/row (lb/row), loam soils, good seedbed. Typical operating speed is 4.7 to 10 km/hr (3 to 6 mi/hr).

Seeding only	450—800/row
	(100—180/row)
Seed, fertilizer herbicides	1100—2000/row
	250—450/row)

Grain Drills

Draft is per furrow opener, and includes wheel rolling resistance. Draft is in Newton per opener — N/opener (lb/opener). Typical operating speed is 3.6 to 10 km/hr (2.5 to 6 mi/hr).

Regular	130—450
	(30—100)
Deep furrow	335—670
	(75—150)

Cultivation

All draft is per unit effective width at typical field speeds. Draft is in Newton per meter — N/m (lb/ft), speed is (S) in kilometers per hour — km/hr (mi/hr), depth (d) in centimeters — cm (in.).

	Draft	**Speed**
Row cultivator	115—230 d	2.5—6.5
	(20—40 d)	(1.5—4)
Lister cultivator	730—2200 d	2.5—6.5
	(50—150 d)	(1.5—4)
Rotary hoe	440 + 21.7 S	9.2—18
	(30 + 2.4 S)	(5.6—11)

Fertilizer and Chemical Application

Anhydrous ammonia applicator	1800 N/knife
	(400 lb/knife)
Fertilizer, pesticide distributors	Rolling resistance only

Rolling resistance is an additional draft force that must be included in computing implement power requirements. Values of rolling resistance depend on transport wheel dimensions, tire pressure, soil type, and soil moisture. Soil moistures are assumed to be less than field capacity for implement operations. Coefficients of rolling resistance are defined in ASAE Standard ASAE S296, Uniform Terminololgy for Traction of Agricultural Tractors, Self-Propelled Implements, and Other Traction and Transport Devices.

Drawbar Power Required[4]

The drawbar power required to pull a tillage tool can be calculated from:

$$kW = \frac{D \ (N/cm^2) \times W \ (cm) \times d \ (cm) \times S \ (km/hr)}{3600} \qquad (1)$$

or

$$hp = \frac{D \ (lb/in^2) \times W \ (in) \times d \ (in) \times S \ (mi/hr)}{375} \qquad (2)$$

where D = specific draft or draft per unit; W = width of tilled zone or number of units; d = depth of tillage; S = forward speed; 1 kW = 1.341 hp; 1 hp = 0.7457 kW.

References are available[1,4-6] to aid in matching tillage tools to appropriate tractors.

Additional Information

For engineers, designers, and researchers who require more detailed information than is provided herein, Table 1 contains a number of references on selected topics.

Table 1
REFERENCES FOR ADDITIONAL
INFORMATION

DRAFT TOOLS

Topic	Ref.
Forces: Analysis of	7—12
Cutting	7, 10, 13, 14
Measurement	3, 9, 15, 16
Soil: Breakup	3, 7, 13, 17
Dynamics	3, 7, 10, 18
Moisture	7, 14, 19, 41
Sliding resistance	3, 7, 19—23
Texture	7, 14, 18, 25
Speed effects	7, 12, 14, 26—29, 41, 42
Modeling	30, 31
Design	7, 12, 27, 28, 32—35
Adjustment	2, 13, 36
Wear	7, 12, 37—40

PTO-DRIVEN TOOLS

Rotary Tillers

Rotary tillers may be used as primary or secondary tillage tools. They consist of a powered rotating shaft, transverse to the direction of travel, equipped with blades that till the soil. Rotary tillers may be used for fall or spring primary tillage, for seedbed preparation, and for chemical incorporation, in combination with chisels or planters. They may also be used as row crop cultivators.

Rotary tillers are versatile machines in that the degree of soil disruption and residue burial can be varied by changing rotary speed, forward speed, depth of operation, blade shape, and shield setting. High rotary speed and a lowered shield result in a fine tilth for incorporation or seedbed preparation, whereas a low rotary speed and raised shield provide a coarser tilth and less residue mixing.

Recommended Power Requirements

Figure 1 illustrates the manufacturers' recommended power requirements per unit of width (kW/m [hp/in.]) for rotary tillers, rotary cultivators, powered harrows,and garden tillers vs. the total width of tillage.

Recommended power requirements for field-size tillers range from 12 kW/m (0.4 hp/in.) to 35 kW/m (1.2 hp/in.). These rotary tillers are more strongly built than rotary cultivators and often have multispeed transmissions. Low power requirements are applicable to tillage of light soils and high power requirements to tillage of heavy soils. Factors other than soil type also affect the required power input:

1. Rotor speed (increasing speed increases power)
2. Rotor diameter (large diameters require more power than small diameters)
3. Type of blade (in order of decreasing power requirement, ''L,'' ''C,'' ''Pick,'' and ''Spike'')
4. Previous soil tillage or soil compaction
5. Soil moisture content
6. Number of blades per flange (increasing number of blades increases power)
7. Ground speed (increasing speed increases power requirement)

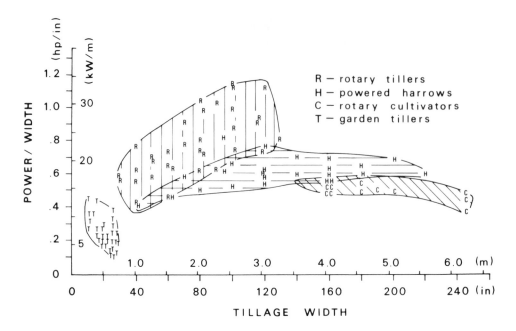

FIGURE 1.   Power requirement per unit width vs. tillage width for PTO-driven tillers and garden tillers, based on manufacturers' recommendations. (NTML Photo No. P-10,307).

8.   Type of residue
9.   Depth of operation

Because of the number of variables that affect the power requirement, and because of their complex interrelationships, there is no accurate method for calculating the power input required. Considerable judgment and experience are required to predict a suitable match among tillage objectives, soil conditions, tiller size, and power requirement. Attempts have been made to correlate power requirements with soil conditions,[1,2] but they have not been verified over a range of field conditions. Two tests[3] in the same soil with the same machine, and in which soil physical measurements were almost identical, resulted in a difference of 50% in power requirements.

When rotary tillers have been used for primary tillage to provide a coarse tilth at a first-pass depth to 15 cm (6 in.), or to provide a fine spring seedbed tilth to 10 cm (4 in.) deep, an average power requirement estimate of 30 kW/m (1 hp/in.) has generally been satisfactory. Light soils may be tilled with 20 kW/m (0.7 hp/in.) or less, whereas heavy soils may require 35 kW/m (1.2 hp/in.). If the intended tillage cannot be accomplished with the manufacturers' recommended power input (overpowering is not recommended because it may lead to machine or drive-line failure), then the depth of tillage or the ground speed must be reduced. Another alternative is to make two passes over the field. References are available[4,5] that outline, in considerable detail, the machine configurations that will produce a variety of final soil conditions. Primary tillage speed generally ranges from 1.5 km/hr (1 mi/hr) to 5 km/hr (3 mi/hr).

Figure 1 includes several field-size tillers with recommended power requirements below 20 kW/m (0.6 hp/in.). Those tillers are generally recommended for use in light soils, or with small tractors at tillage depths or speeds lower than those normally used for field operations. The field-size tillers can also be used for secondary tillage operations.

Rotary Cultivators

The rotary cultivators represented in Figure 1 were intended for use in secondary tillage

Table 2
ADDITIONAL ROTARY TILLAGE REFERENCES

Topic	Ref.
Blades: Arrangement	2, 12, 13
Design	2, 7, 12, 14—16
Draft	2, 6, 12, 14, 18
Drive-lines	3, 19—21
Incorporation	6, 17, 22, 23
Multipowered tools	2, 7, 24—28
Power requirements	2, 3, 7, 12, 17, 24, 29—33, 40
Pulverization	2, 6, 17, 24, 34, 35
Seedbed preparation	12, 17, 36—39
Speed (rotary vs. forward)	2, 7, 12, 17, 24, 34, 41

operations. The recommended power input of 14 kW/m (0.4 hp/in.) to 18 kW/m (0.6 hp/in.) is comparable to that for small rotary tillers. The speed range for rotary cultivation is normally 5.6 km/hr (3.5 mi/hr) to 8 km/hr (5 mi/hr).

Rotary Tiller Draft Requirements

One of the advantages of rotary tillers is that the power is transmitted by a drive-line instead of by developing drawbar pull via the soil-tire interface. Because of the method of power transmission, conventional rotary tillers tend to develop a pushing force on the tractor and, thereby, help move the machine forward. In most instances, the forward thrust is small (equal to approximately 10% of the power transmitted to the rotor),[6] but, in conditions in which large, slow-turning rotors take large increments of cut, as much as 30%[7] of the power input to the rotor may be recovered as thrust on the drawbar. When drawbar thrusts are large, utilization of that thrust to pull chisels or other ground-engaging implements is desirable; continued large drawbar thrusts may be injurious to the tractor transmission and drive-line.[8]

Powered Harrows

Powered harrows are equipped with horizontally rotating, oscillating, or reciprocating tines or spikes. They are used mainly for secondary tillage, but have been used on a limited basis as primary tillers for light soils. Figure 1 indicates that the recommended power requirement for powered harrows is between 12 kW/m (0.4 hp/in.) and 23 kW/m (0.8 hp/in.).[9] The normal operating speed of powered harrows is from 4.7 kW/m (3 mi/hr) to 13 kW/m (8 mi/hr).

Garden Tillers

The range of recommended power for garden tillers is presented in Figure 1, chiefly as a matter of interest. The low power/width and high power/width values were obtained with the same tillers, but additional sections of tines were added to the tillers for the low power/width ratios to make them suitable as garden cultivators.

Additional Information

For engineers, designers, and researchers who require more detailed information than is provided herein, two annotated bibliographies[10,11] are available that contain over 800 references on all aspects of rotary tillage tools. Table 2 contains a limited number of references on selected topics.

ACKNOWLEDGMENT

The author expresses appreciation for the suggestions provided by G. D. Tedder, Vice-President, Engineering, Howard Rotavator Co., Inc.

REFERENCES

Draft Tools

1. American Society of Agricultural Engineers, Agricultural Machinery Management Data (ASAE D230.3 revised Dec. 1977), *Agricultural Engineers' Yearbook,* American Society of Agricultural Engineers, St. Joseph, Mich., 1979, 248.
2. **Buckingham, F.,** Fundamentals of Machine Operation: Tillage, John Deere Serv. Publ., John Deere Co., Moline, Ill., 1976.
3. **Kepner, R. A., Bainer, R., and Barger, E. L.,** *Principles of Farm Machinery,* AVI Publishing, Westport, Conn., 1972.
4. **Bowers, W.,** Matching Equipment to Big Tractors for Efficient Field Operations, ASAE Tech. Paper No. 78-1031, American Society of Agricultural Engineers, St. Joseph, Mich., 1978.
5. **Bowers, W.,** Matching Tillage Implements to Big Tractors, Okla. State Univ. Ext. Facts No. 1209, Oklahoma State University, Stillwater, 1978.
6. **J. I. Case Co.,** A Model for Tractor/Implement Selection, Bulletin, Rev. No. 238OMTIS, J. I. Case Agricultural Equipment Division, Racine, Wis.
7. **Gill, W. R. and Vanden Berg, G. E.,** Soil Dynamics in Tillage and Traction, Agriculture Handbook No. 316, U.S. Department of Agriculture, Washington, D.C., 1967.
8. **Vanden Berg, G. E.,** Analysis of forces on tillage tools, *J. Agric. Eng. Res.,* 11, 201, 1966.
9. **Clyde, A. W.,** Force measurements applied to tillage tools, *Trans. ASAE,* 4, 153, 1961.
10. **Soehne, W.,** Some principles of soil mechanics as applied to agricultural engineering, *Grundlagen Landtechnik,* (NIAE transl. No. 53), 7, 11, 1956.
11. **Morling, R. W.,** Soil Force Analysis as Applied to Tillage Equipment, ASAE Tech. Paper No. 63-149, American Society of Agricultural Engineers, St. Joseph, Mich., 1963.
12. **Richey, C. B., Jacobson, P., and Hall, C. W.,** *Agricultural Engineers' Handbook,* McGraw-Hill, New York, 1961.
13. **Gill, W. R. and McCreery, W. F.,** Relation of size of cut to tillage tool efficiency, *Agric. Eng.,* 41, 372, 1960.
14. **Randolph, J. W. and Reed, I. F.,** Tests of tillage tools. II. Effects of several factors on the reactions of fourteen-inch moldboard plows, *Agric. Eng.,* 19, 29, 1938.
15. **Zoerb, G. C.,** A strain gage dynamometer for direct horsepower indication, *Agric. Eng.,* 44, 434, 1963.
16. **Hendrick, J. G.,** 3-Point hitch dynamometer, in section Powered Tillage Tools, Annual Report, National Tillage Machinery Laboratory, Auburn, Ala., 1973.
17. **Bateman, H. P., Naik, M. P., and Yoerger, R. R.,** Energy required to pulverize soil at different degrees of compaction, *J. Agric. Eng. Res.,* 10, 132, 1965.
18. **Kisu, M.,** Soil physical properties and machine performances, *Jpn. Agric. Res. Q.,* 6, 151, 1972.
19. **Nichols, M. L.,** The dynamic properties of soil. II. Soil and metal friction, *Agric. Eng.,* 12, 321, 1931.
20. **Cooper, A. W. and McCreery, W. F.,** Plastic Surfaces for Tillage Tools, ASAE Tech. Paper No. 61-649, American Society of Agricultural Engineers, St. Joseph, Mich., 1961.
21. **Soehne, W.,** Friction and cohesion in arable soils, *Grundlagen Landtechnik,* (transl. by Dept. of Irrigation and Soils, University of California, Los Angeles), 5, 64, 1953.
22. **Schafer, R. L., Gill, W. R., and Reaves, C. A.,** Experiences with lubricated plows, *Trans. ASAE,* 22, 7, 1979.
23. **Fox, W. R. and Bockhop, C. W.,** Characteristics of a Teflon-covered simple tillage tool, *Trans. ASAE,* 8, 227, 1965.
24. **Dano, P. L. and Gill, W. R.,** The influence of electro-osmosis and polytetrafluorethylene on reducing sliding friction in soils, in *Proc. Int. Conf. Energy Conserv. Crop Prod.,* Massey University, New Zealand, 1978, 160.
25. **Payne, P. C. J.,** The relationship between the mechanical properties of soil and the performance of simple cultivation implements, *J. Agric. Eng. Res.,* 1, 23, 1956.

26. **Rowe, J. R. and Barnes, K. K.,** Influence of speed on elements of draft of a tillage tool, *Trans. ASAE,* 4, 55, 1961.
27. **Soehne, W.,** Investigations on the shape of plough bodies for high speeds, *Grundlagen Landtechnik,* (NIAE transl. No. 87), 11, 22, 1959.
28. **Bernacki, H., Haman, J., and Kanafojski, Cz.,** *Agricultural Machines, Theory and Practice,* Vol. 1, U.S. Department of Agriculture and National Science Foundation, Washington, D.C., 1972.
29. **Olson, D. J. and Weber, J. A.,** Effect of Speed on Soil Failure Patterns in Front of Model Tillage Tools, SAE Tech. Paper No. 650691, Society of Automotive Engineers, New York, 1965.
30. American Society of Agricultural Engineers, *Similitude of Soil Machine Systems,* ASAE Publ. No. 3-77, American Society of Agricultural Engineers, St. Joseph, Mich., 1977.
31. **Barnes, K. K., Bockhop, C. W., and McLeod, H. E.,** Similitude in studies of tillage implement forces, *Agric. Eng.,* 41, 32, 1960.
32. **Sineokov, G. N.,** *Design of Soil Tilling Machines,* U.S. Department of Agriculture and National Science Foundation, Washington, D.C., 1965.
33. **Nichols, M. L. and Reaves, C. A.,** Soil reaction: to subsoiling equipment, *Agric. Eng.,* 39, 340, 1958.
34. **Nichols, M. L., Reed, I. F., and Reaves, C. A.,** Soil reaction: to plowshare design, *Agric. Eng.,* 39, 336, 1958.
35. **Tupper, G. R. and Barrentine, W. L.,** Design of the Stoneville Parabolic Super Chisel, Miss. Agric. and Forest Exp. Stn. Inf. Sheet No. 1291, 1979.
36. **Reed, I. F.,** Disk Plows, Farmers' Bull. No. 2121, U.S. Department of Agriculture, Washington, D.C., 1964.
37. **Richardson, R. C. D.,** The wear of metallic materials by soil-practical phenomena, *J. Agric. Eng. Res.,* 12, 22, 1967.
38. **Moore, M. A.,** The abrasive wear resistance of surface coatings, *J. Agric. Eng. Res.,* 20, 167, 1975.
39. American Society for Testing Materials, Standard practice for the dry sand/rubber wheel abrasion test. ASTM Standard (in process of approval), American Society for Testing Materials, Philadelphia, Pa., 1980.
40. **Reed, I. F. and Gordon, E. D.,** Determining the relative wear resistance of metals, *Agric. Eng.,* 32, 98, 1951.
41. **Vershinin, P. V. and Kosarchuk, A. K.,** The influence of moisture on draft during high speed plowing, (transl. available from Natl. Tillage Machy. Lab., Auburn, Ala., *Sel'khoz. Nauki. USSR* 61, Feb., 1960.
42. **Schlegel, J. E. and Morling, R. W.,** Optimum travel speed for maximum plowing acreage, *Trans. ASAE,* 12, 690, 1969.

PTO-Driven Tools

43. **Perdok, U. D. and Burema, H. J.,** Power Requirements of Rotary Tiller Blades in Clay and Sandy Soils, Res. Rep. 771, Institute of Agricultural Engineering, Wageningen, The Netherlands, 1977.
44. **Bernacki, H., Haman, J., and Kanafojski, Cz.,** Agricultural Machines, Theory and Construction, Vol. 1, U.S. Department of Agriculture and National Science Foundation, Washington, D.C., 1972.
45. **Richardson, R. D.,** Some torque measurements taken on a rotary cultivator, *J. Agric. Eng. Res.,* 3, 66, 1958.
46. **Buckingham, F.,** Fundamentals of Machine Operation: Tillage, John Deere Serv. Publ., John Deere Co., Moline, Ill., 1976.
47. **Howard Rotavator Co.,** Pocket Guide to Rotavation, Form No, 514, Howard Rotavator Co., Harvard, Ill.
48. **Adams, W. J. and Furlong, D. B.,** Rotary tiller in soil preparation, *Agric. Eng.,* p. 600, 1959.
49. **Chamen, W. C. T., Cope, R. E., and Patterson, D. E.,** Development and performance of a high output rotary digger, *J. Agric. Eng. Res.,* 24, 301, 1979.
50. **Wismer, R. D., Wegscheid, E. L., Luth, H. J., and Romig, B. E.,** Energy Application in Tillage and Earthmoving, Soc. Automot. Eng. Paper, FCIM Mtg., Society of Automotive Engineers, Milwaukee, Wis., September 1968.
51. **Ewing, C. G. and Butterworth, B.,** Scanning the power harrows, *Implement Tractor,* 94, 34, 1979.
52. **Hendrick, J. G.,** An Annotated Bibliography on Rotary Tillage Tools, National Tillage Machinery Laboratory, U.S. Department of Agriculture, Auburn, Ala., 1970.
53. **Hendrick, J. G. and Bailey, A. C.,** An Annotated Bibliography on Rotary Tillage Tools, Suppl. No. 1, National Tillage Machinery Laboratory, U.S. Department of Agriculture, Auburn, Ala., 1978.
54. **Dalin, A. D. and Pavlov, P. V.,** Rotary Soil Cultivating and Excavating Machines, U.S. Department of Agriculture and National Science Foundation, Washington, D.C., 1970.
55. **Soehne, W.,** Influence of shape and arrangement of tools on torque of rotary blades, (Unnumbered NIAE transl.), *Grundlagen Landtechnik,* p. 69, 1975.
56. **Beeny, J. M. and Khoo, D. C. P.,** Preliminary investigations into the performance of different shaped blades for the rotary tillage of wet rice soil, *J. Agric. Eng. Res.,* 15, 27, 1970.

57. **Hendrick, J. G. and Gill, W. R.,** Rotary tiller design parameters. IV. Blade clearance angle, *Trans. ASAE,* 17, 4, 1974.

58. **Tsuchiya, M.,** Studies on Power Tillers in Japan, (transl. by Shin-Norin Co., Ltd.), Yamagata University, Tsuruokashi (Japan), 1965, 1.

59. **Furlong, D. B.,** Rotary Tiller Performance Tests on Existing Tines, Tech. Rep. No. 1049, FMC Corp., San Jose, Calif., 1956.

60. **Hendrick, J. G. and Gill, W. R.,** Rotary tiller design parameters. I. Direction of rotation, *Trans. ASAE,* 14, 669, 1971.

61. **Crolla, D. A. and Chestney, A. A. W.,** Field measurements of driveline torques imposed on PTO driven machinery, *J. Agric. Eng. Res.,* 24, 157, 1979.

62. **Hilton, D. J. and Chestney, A. A. W.,** The reduction of torsional vibration in a rotary cultivator transmission by means of resilient coupling, *J. Agric. Eng. Res.,* 18, 47, 1973.

63. American Society of Agricultural Engineers, Operating requirements for power-take-off drives, ASAE Standard 207.9 (SAE J 721 f), *Agricultural Engineers' Yearbook,* American Society of Agricultural Engineers, St. Joseph, Mich., 1979—80, 158.

64. **Bode, L. E. and Gebhardt, M. R.,** Equipment for incorporation of herbicides, *Weed Sci.,* 17, 551, 1969.

65. **Carter, L. M. and Miller, J. H.,** Characteristics of powered rotary cultivators for application of herbicides, *Trans. ASAE,* 12, 5, 1969.

66. **Sineokov, G. N. and Panov, I. M.,** Share plows with combined working tools, in Theory and Calculation of Soil Working Machines, U.S. Department of Agriculture and National Science Foundation, Washington, D.C., 1978, chap. 9.

67. **Yuzbashev, V.,** The influence of preliminary loosening on the torque of a rotary plow, Report, *All-Union Sci. Res. Inst. Mach. Const. USSR,* 70, 142, 1972.

68. **Panov, I. M., Petrov, S. N., Melikhov, V. V., and Yuzbashev, V. A.,** Decrease of energy requirement of a rotary plow, *Mech. Elec. Socialist Agric.,* USSR, 2, 20, 1971.

69. **Toma, D.,** PTO-Driven Machines for the Rotary Cultivation of the Soil, AGRI/MECH/39, United Nations, New York, 1969.

70. **Soane, B. D.,** Review of cultivation and compaction research, Proc. Subject Day on Mechanical Behavior of Agricultural Soils, NIAE Rep., No. 7, Paper No. 2, June 1973.

71. **Billot, J. F. and Binesse, M.,** Some measurements on PTO-driven implements for soil cultivation (NIAE transl. No. 318), *Bull. Inf. CNEEMA,* France, 174/175, 33, 1972.

72. **Ghosh, B. N.,** The power requirement of a rotary cultivator, *J. Agric. Eng. Res.,* 12, 5, 1967.

73. **Hendrick, J. G. and Gill, W. R.,** Rotary tiller design parameters. II. Depth of tillage, *Trans. ASAE,* 14, 675, 1971.

74. **Sineokov, G. N.,** Design of Soil Tilling Machines, U.S. Department of Agriculture and National Science Foundation, Washington, D.C., 1977.

75. **Beeny, J. M. and Greig, D. J.,** The efficiency of a rotary cultivator, *J. Agric. Eng. Res.,* 10, 5, 1965.

76. **Hendrick, J. G. and Gill, W. R.,** Rotary tiller design parameters. III. Ratio of peripheral and forward velocities, *Trans. ASAE,* 14, 679, 1971.

77. **Frevert, R. K.,** Mechanics of Tillage, M.S. thesis, Iowa State University, Ames, 1940.

78. **Dalleinne, E.,** Minimum tillage with rotary cultivators equipped with horizontal rotors (NIAE transl. no. 311), *Bull Inf. CNEEMA,* France, 165, 29, 1971.

79. **Griffith, D. R. et al.,** Effect of eight tillage-planting systems on soil temperature, percent stand, plant growth, and yield of corn on five Indiana soils, *Agron. J.,* 65, 321, 1973.

80. **Hoyle, B. J. and Yamada, H.,** Seedbeds and Aggregate Stability Improved by Rototilling Wet Soil, Soil Sci. Soc. Am. Spec. Publ. Ser. No. 7, Soil Science Society of America, Madison, Wis., 1975, 111.

81. **Smith, E. M. and Benock, G.,** Power Tillage Compared with Conventional Tools for Grassland Renovation, Am. Soc. Agric. Eng. Paper No. 77-1003, American Society of Agricultural Engineers, St. Joseph, Mich., June 1977.

82. **Kisu, M., Kohda, Y., Yagi, S., and Seyama, K.,** Studies on Trafficability, Tractive and Rotary Tilling Performance of Tractor, Tech. Rep., Institute of Agricultural Machinery, Omiya, Japan, 1966.

83. **Beeny, J. M.,** Rotary cultivation of wet rice land — comparison of blade shape, *J. Agric. Eng. Res.,* 18, 249, 1973.

CUTTING OF BIOLOGICAL MATERIALS

William J. Chancellor

PRINCIPLES OF CUTTING

Biological materials commonly subjected to cutting can be classified into two general categories:

1. Nonfiberous materials having uniform properties in all directions; at the time of cutting, the cells of these materials are usually turgid with liquid cell materials.
2. Fiberous materials with high tensile strength fibers oriented in a common direction and with comparatively low-strength materials bonding the fibers together.

It is believed that with the first category of materials the concentrated compressive stress applied by the cutting tool to the cell wall causes cell pressure to increase and produce a tensile failure in the cell wall at the point of contact with the cutting tool.[1] This point is subjected to:

1. Accentuated tensile stresses due to the addition of bending stresses at the point of indentation, to the hydrostatically caused stress,
2. A uniquely high shear stress because of the applied compression stresses in one direction and the hydrostatically caused tensile stresses in an orthogonal direction,

$$\text{Shear Stress} = \frac{\text{Tensile Stress} + \text{Orthogonal Compressive Stress}}{2}$$

3. And movement of the tool in a direction parallel to its edge and perpendicular to the direction of the compressive force applied (a slicing action) which can further add to the shear stress applied to the cell wall at the point of contact.

When the blade is sharp and hydrostatic cell pressure is high, the blade force necessary to produce cell wall failure is sufficiently low to prevent much compressive deformation of the material at any point other than that of the cell wall in immediate contact with the blade. The resulting continuous or gliding type of cut has also been observed when making cross-grain cuts of very turgid fiberous material using a very sharp blade.[1]

Generally, cross-grain cuts of fiberous material (particularly that which is not turgid) involve the cutting tool first compressing the fibers against:

1. A support which is part of the cutting mechanism,
2. Structural support from the fiber mass itself (as in the case of a block of wood), or
3. Inertial forces caused by attempting to accelerate the mass of the material attached to the point of blade-material contact.

As this compression (in a direction perpendicular to the fibers) proceeds, the shear strength of the material is mobilized to produce tensile stresses in the fibers. Finally these stresses become so great that the fibers fail in tension.[1,2] Ultimate tensile stresses and corresponding shear stresses at failure for alfalfa and corn stems have been studied.[3,4]*

---

* Ultimate tensile strength of alfalfa stems was approximately 8.87 N for a stem having 0.00086 g dry matter per centimeter of length, and corresponding ultimate shear strength was 3.26 N for the same-sized stem. Ultimate tensile strength of corn stalks was approximately 7015 N/cm² (about 27,148 N for a stem having 1.32 g of dry matter per centimeter of length) for the lower internodes irrespective of moisture content.

The compressive stresses required to produce such a failure:

1.   Require such high forces that the material is noticeably compressed and flattened before the fibers immediately adjacent to the blade begin to fail, and
2.   Can be achieved with low blade forces if a high degree of stress concentration is produced by firm, concentrated counter-cutting support (thick masses of material tend to diffuse this stress concentration).

The compressive stress that must be applied by the blade edge to cause cutting (σ_u) is believed to be a characteristic of the material.[5] For common crop materials, values of σ_u range from 8.8×10^6 to 29.9×10^6 N/m². The blade force, per unit blade length, to cause cutting (P_{cr}) has been hypothesized[5] to be representable by:

$$P_{cr} = \delta\sigma_u + \frac{C}{2}\frac{(h_c)^2}{E}\ [\tan B + f\sin^2 B + \mu(f + \cos^2 B)]$$

in which: δ = the width of the blade edge (m), C = the compressive modulus of the material (N/m²), E = the free-standing depth of the material layer (m) (Figure 1), h_c = the depth of the compressed material layer just as cutting is beginning (m) (Figure 1), B = the bevel angle of the blade edge (degrees), f = the coefficient of friction between the blade and the material to be cut, and μ = Poisson's ratio of the material being cut.

When fiberous materials are cut in such a way that the individual fibers remain intact, but become separated from each other as a result of cutting action, groups of fibers are initially separated by blade cleavage. Beam-like characteristics of these groups of fibers permit the tearing apart (tensilly) of the soft materials bonding the fibers together (splitting). This failure may extend some distance ahead of the blade.[2,6]

PARAMETERS OF CUTTING PERFORMANCE

Forces

The forces on cutting tools are important factors for cutting equipment designers. Two main force parameters have been used:

1.   The force in the direction of relative tool movement, per unit of projected cutting width, and
2.   The force in the direction of relative tool movement, per unit of cross-sectional area of the material being cut.

With biological materials, it may be difficult to define an exact cross-sectional area, and even if this is accomplished, that area may contain a sizable proportion of air-filled or liquid-filled pores. To overcome these problems, the cross-sectional area may be specified in terms of the mass of dry matter (d.m.) per unit length of the assemblage being cut. This is the same as normalizing the force per unit area by dividing by the d.m. density of the material being cut. Thus the force per unit cross-sectional area of material cut would be in terms of N/(g d.m./cm) = (N/cm²)/(g d.m./cm³) = N·cm/g d.m. Forces acting in directions normal to that of relative tool movement may also be of interest, as they affect penetration, control, and support forces for blades and cutters.

Energy Requirements

Energy consumption in cutting is an important design factor. It is sometimes reported in terms of the energy consumed per unit of area cleaved by the cutting process. If cross-

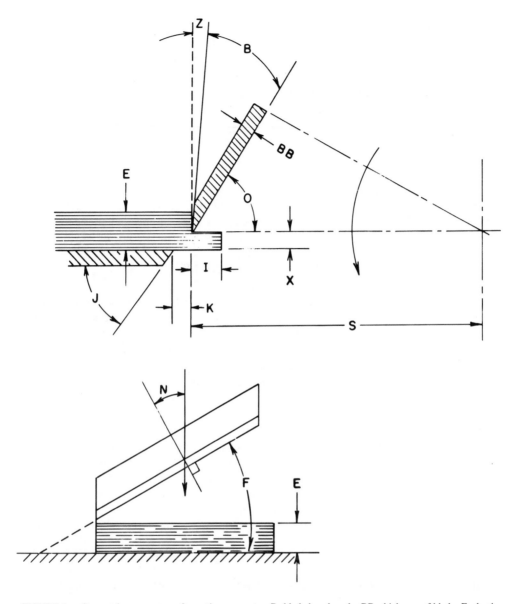

FIGURE 1. Geometric parameters of a cutting apparatus. B, blade bevel angle; BB, thickness of blade; E, depth of material cut; F, included angle between edges; I, length of material cut off from main piece; J, shear bar clearance angle; K, clearance between cutting edges; N, angle of normal to blade edge, to direction of blade motion; O, rake angle; S, edge path radius; X, shear bar position relative to center plane of blade cylinder; Z, back of blade clearance angle.

sectional area is specified in terms of the mass of d.m. per unit length of the assemblage being cut, then the energy consumed per unit cross-sectional area cleaved can be expressed as energy per unit d.m. mass cut to a given length. This term is merely the ratio of the energy per unit cross-sectional area and the d.m. density of the material. Both these components may vary during the cutting process, but their ratio generally represents a stable parameter for a given cutting operation, $(kWhr/cm^2)/(t\ d.m./cm^3) = \dfrac{kWhr \cdot cm}{t\ d.m.}$. This term may be applied directly to forage harvester operations:

$$\frac{\text{kWhr·cm}}{\text{t d.m.}} \times \frac{\text{t d.m.}}{\text{h}} \times \frac{1}{\text{cut length (cm)}} = \text{kW}$$

For mower operations, it may be assumed that the plant cross-sectional area at the cutting height is approximately twice the average cross-sectional area of the crop. Thus:

$$\frac{\text{kWhr·cm}}{\text{t d.m.}} \times \frac{\text{t d.m.}}{\text{ha}} \times 2 \times \frac{\text{ha}}{\text{hr}} \times \frac{1}{\text{crop height (cm)}} = \text{kW}$$

Other Parameters

Among the other parameters occasionally of concern are items such as (1) quality of cutting (degree of fraying of particles cut), (2) proportion of stems cut, (3) velocity and direction with which cut materials are pushed before cutting and impelled after cutting, (4) wear caused on cutting parts, and (5) shatter of plant parts such as seeds or leaves.

CUTTING MECHANISMS

There are many different mechanisms used for cutting biological materials. Detailed attention will be given here to three of the most common: (1) forage harvester shear-bar type cutterhead, (2) hay mower cutter-bar, and (3) flail-type mower. Because these mechanisms are difficult to monitor in detail during field operations, much of the information available on cutting performance was developed in the laboratory — both with equipment designed to simulate the field-machine cutting process, and with controlled laboratory tests with the field machine itself. Data from field tests are also available.

Forage Harvester Shear-Bar Type Cutterhead

The process involved here consists of cutting groups of plant stems on a plane perpendicular to the fiber direction, by a knife moving past a fixed, flat shear bar. Many researchers have worked to determine cutting performance parameters for this sort of mechanism. These are summarized in the following Table 1 for the two most commonly cut materials, alfalfa *(Medicago sativa)* stems and corn *(Zea mays)* stalks.

Averages of the lowest and highest cutting-energy values (exclusive of those including additional factors) from each researcher were, respectively, 1.08 and 1.96 $\frac{\text{kWhr·cm}}{\text{t d.m.}}$ for alfalfa, and 1.04 and 2.46 $\frac{\text{kWhr·cm}}{\text{t d.m.}}$ for corn. These values indicate there was little difference between these two materials and that the cutting process variables investigated by each researcher tended to have a limited range of influence on cutting energy requirements.

The forage harvester cutterhead frequently has a number of other functions besides cutting. Some are air movement, particle throwing, and particle grinding. Energy requirements for these functions in field machines generally exceeded that for cutting.

Hay Mower Cutter Bar

With this cutting mechanism, stems supported at ground level are cut more or less individually between a sharpened, reciprocating blade and a supporting edge (ledger plate), which is also usually beveled and frequently serrated. Design considerations related to the reciprocating drive are frequently given first priority, and factors related to cutting are mainly concerned with effectiveness rather than force and power requirements.

With hay-mower-type machines, the cutting force and energy values frequently appeared several times larger in measurements with actual field machines, than in measurements with

Table 1
PERFORMANCE PARAMETERS — FORAGE HARVESTER, SHEAR-BAR
TYPE MACHINES

Test type[a]	Material	Moisture content percent (w.b.)	Cutting energy kWhr·cm / tonne d.m.	Cutting force N·cm / g d.m.	Cutting force N / cm	Ref.
L-L	Alfalfa	20	0.807—1.36	340—761	92—165	1
L-F	Alfalfa	74	3.35—4.49 (includes air movement energy)			7
F-F	Corn	45—70	0.83—2.70			8
F-F	Corn	75	0.998—2.16			9
F-F	Alfalfa	72	1.28—2.49			
L-L	Alfalfa	28—60	0.56—1.58			10
L-L	Timothy	7—72	1.86—2.05			11
L-L	Alfalfa	43	1.41—1.46	985—1257	306—420	12
L-F	Alfalfa	58	1.25—1.46			13
L-F	Corn	64	1.65—1.78			
L-F	Alfalfa	42—69	0.93—2.13	392—387		14
L-L	Alfalfa	5—28	1.79—4.06			15
L-F	Corn	71	3.10—4.83 (includes accelerating energy)			16
L-F	Alfalfa	70	0.17—0.91			17
L-F	Corn stalks	19—27	0.69—1.41			18
	Corn stalks and ears	22	0.41			
L-L	Corn	75—80	1.26—2.49	70—210		19
L-F	Alfalfa	6—10	1.29—1.92			20
F-F	Alfalfa	56	1.32—2.20			21, 22
F-F	Corn	60	0.86—4.21			

[a]   L-L indicates laboratory tests with laboratory apparatus, L-F — laboratory tests with a field machine, F-F — field tests with a field machine.

laboratory equipment. Researchers who observed this point hypothesized that stalk bending, cutting inclined stalks, cutting one stalk against another, as well as increased machine friction due to the presence of material being cut, were likely sources responsible for much of the difference.[23,24] The data presented in Table 2 indicate that with this type of machine the cross-sectional area of the material cut by each blade (as measured by d.m. mass per unit length) was not a particularly useful parameter in normalizing the cutting force values. This was because the stalks are not generally confined during cutting and can be cut individually (low force values) or can be clumped together to cause high force values. However, when clumped, the entire mass does not fail simultaneously, and consequently the force, while high, is not in proportion to the total cross-sectional area cut.

The cutting energy values found in the laboratory with hay-mower-type machines were generally slightly lower than those found for forage harvester shear-bar type machines.

Flail-Type Forage Cutting Machines

This type of device has been used extensively since the early 1950s because of its simplicity, freedom from blockages, and adaptability to widely varying conditions. The principle of operation is that the blades strike the plant at such a high velocity that inertial forces of the plant, as they resist rapid acceleration, provide sufficient resistance for the blade to

Table 2
PERFORMANCE PARMETERS — HAY-MOWER-TYPE MACHINES

Test type[a]	Material	Moisture content percent (w.b.)	Cutting energy kWhr·cm / t d.m.	Cutting force		
				Maximum, N	(N·cm)/g d.m.	Ref.
F-F	Mixed hay	75	5.81	25.43	740.5	25
F-F	Alfalfa	77	16.3—25.9	108—170.7[b]	1886—2987[b]	26
F-F	Corn	86.6	37.4—56.8	148—312[b]	4307—5116[b]	
L-F	Corn	75	0.58	256	167	27
L-F	Rice-straw	71	1.77	280	424	30
L-F	Rice-straw	14	1.37	209	317	
L-L	Timothy	54	0.44	23.6	244	1
L-L	Alfalfa	15	1.20	81	338	
L-L	Alfalfa	77	0.52	26.7	1051	29
L-L	Corn	82	0.62	48.5	248	
L-L	Rice straw	44—65	1.18	14.4	3673	30
L-L	Perennial Ryegrass	75	0.44		44	31

[a] L-L indicates laboratory tests with laboratory apparatus. L-F — laboratory tests with field machine components. F-F — field tests with a field machine.
[b] Average forces based on 41% of stroke with cutting.[32]

generate stresses high enough to cause failure, long before the material reaches blade velocity. High-speed photography has shown that severance of forage stalks takes place with only very minor displacement of the stalk in the immediate zone of the cut.[1,33-35] However, the impulse transmitted during the cutting process results in a high velocity being imparted to the free materials. This velocity is further augmented by continuing contact with the high velocity blade and associated parts.

A theoretical analysis of the critical (minimum) blade velocity necessary to cut a free-standing stalk[35] resulted in the relationship:

$$\text{Critical Velocity (m/sec)} = K \sqrt{\frac{\text{stalk diam. (m)} \times \text{max. cutting force (N)}}{\text{effective stalk mass (kg)}}}$$

K = 1 when cutting force increases linearly from zero at the beginning of the cut to a maximum value at the end of the cut; K = 1.41 when cutting force is uniform from the beginning to the end of the cut. The effective stalk mass was found to be dependent on: (1) stalk stiffness (product of modulus of elasticity and the axial moment of inertia of the cross-section surface), (2) stalk mass per unit length, and (3) stalk acceleration at the point of cutting.

Since stalk stiffness is a function of (stalk diameter),[4] mass per unit length a function of (stalk diameter),[2] and the maximum force a function of (stalk diameter),[1] the following relations were formulated[35] using typical plant parameters:

1. For the case where maximum cutting force (N) = 19.62 (stalk diameter, mm), critical blade velocity (m/sec) = 62 − 1.2 (stalk diameter, mm)
2. For the case where maximum cutting force (N) = 9.81 (stalk diameter, mm), critical blade velocity (m/sec) = 47 − 1.3 (stalk diameter, mm)
3. For the case where maximum cutting force (N) = 4.905 (stalk diameter, mm), critical blade velocity (m/sec) = 32 − 0.75 (stalk diameter, mm)

Table 3
PERFORMANCE PARAMETERS — FLAIL-TYPE FORAGE CUTTING MACHINES

Material	Moisture content percent (w.b.)	Velocity (m/sec)	Cutting height (cm)	Cutting energy kWhr·cm / t d.m.	Force N/cm	Type[a]	Ref.
Sunflower	81	7—26	5	8.71	73[b]	M,L-L	33—35
Alfalfa	54	11—26		9.2	195[b]	M,L-L	35
Wheat	47.5	5—26			57		
Rape	78	7—26		9.2	63		
Alfalfa	75	2.5—9.6	0.6—7.6	2.63—3.63		M,L-L	36
Sudan grass				3.03—4.68			
Grasses	74	20—60	2.0	20.98	79	M,L-L	37
Oats				18.4			
Grasses	74	50—100	3.2	39.24		M,F-F	38
Alfalfa		10—40	3.8	3.1—12.4		M,L-L	29
Oats				3.6—7.2			
Corn	87	60	3.0	100.9		M,F-F	26
Alfalfa	77			52.3			
Douglas fir	5	58—117		6.43		C,L-F	39
Lodgepole pine	27			4.27			
Alfalfa	73	66		24.39		C,F-F	40
Sudan grass	70			17.03			
Alfalfa	63	56	1.00—17.5	30.56		C,F-F	41
Red clover	70			25.38			
Corn	83	30—49		52.44		C,F-F	42
Soybeans	81			36.09			

[a]   M stands for Mower, C for Chopper, L-L is a laboratory test with a laboratory machine, L-F is a laboratory test with field machine components, F-F is a field test with a field machine.
[b]   For the four materials involved, the cutting force data were as follows:

		N/stem		N·cm
	N/cm plant diam.	Range	Typical	g d.m.
Sunflower	73	78—127	88	545
Rape	63	39—69	44	333
Alfalfa	195	17—157	88	383
Wheat	57	10—49	20	952

1.   For the case where maximum cutting force (N) = 19.62 (stalk diameter, mm), critical blade velocity (m/sec) = 62 − 1.2 (stalk diameter, mm)
2.   For the case where maximum cutting force (N) = 9.81 (stalk diameter, mm), critical blade velocity (m/sec) = 47 − 1.3 (stalk diameter, mm)
3.   For the case where maximum cutting force (N) = 4.905 (stalk diameter, mm), critical blade velocity (m/sec) = 32 − 0.75 (stalk diameter, mm)

The flail-type cutting process requires more energy, both in the field and in the laboratory, than corresponding processes with conventional mowers or shear-bar type forage choppers. Comparison of data in Table 3 with those in Tables 1 and 2 indicates ratios of energy requirement increase from two to ten times.

With flail-type forage cutting machines as with conventional mowers, the energy requirements of field machines appear to be several times greater than those found for the same type of cutting process in the laboratory. In the case of flail-type machines, much energy is used for accelerating the cut material and pumping air. With the flail chopper, reduction of particle length by beating and tearing rather than by cutting causes high energy requirements. Furthermore, these choppers frequently move the material along an enclosing hood to a discharge point that induces friction energy requirements.

Blade forces at the high velocities found in flail-type cutting devices are difficult to measure accurately. Consequently, such data were reported in only two studies, and these were obtained with elaborate instrumentation.[33-35,37,38] These force values, however, were found to be in approximately the same range as those found with conventional mowers.

Blade velocities of at least 10 m/sec were required for effective cutting and minimum values of 20 to 25 m/sec were needed in some cases. Lower blade velocities could be tolerated as the height of cut above the point of plant support was reduced.

Other Cutting Mechanisms

The literature shows an extensive number of other cutting mechanisms. These are reviewed as follows:

1. *Machete*. Typical performance of a man using a machete to cut sugar cane was 0.55 ton of cane per hour.[43]
2. *Sickle*. Average values from several studies of cutting rice with a sickle showed 79 man-hours per hectare to cut a crop consisting of about 2000 kg grain and 2000 kg straw per hectare.[44] Typical maximum forces for cutting rice straw with both serrated and smooth sickles was 103 N (per bundle) when cutting a bundle of 12 stems, each bundle having 0.56 g of d.m. per centimeter of length at cutting height, resulting in a figure of $(183 \frac{\text{N·cm}}{\text{g d.m.}})$. Energy requirements were $2.87 \frac{\text{kWhr·cm}}{\text{t d.m.}}$ for both types of sickles. This was about 150% of the energy requirement found using a conventional mower for similar stalks.[28]
3. *Double-Knife Reciprocating Mower*. An analysis of the action of such a mower showed that it cut stems with less crowding and distributed cutting action over a greater proportion of the knife stroke than was the case with a conventional single-knife mower. Problems with stalks being forced forward from between the blades were related to knife geometry and knife-plant friction.[32] Tests of such a mower showed that for most crops energy requirements were higher (75% in the case of young corn) than when using a conventional cutter bar. For alfalfa and ladino clover, energy requirements for the two mower types were similar.[26]
4. *High-Speed Band or Belt Cutters*. A V-belt with "C-section" width and "B-section" depth with metal cutters attached was investigated as a possible vegetation cutting device. A belt speed of 40 m/sec was used and a reel-type device was found necessary to move the forage over the 30 to 38 cm longitudinal dimension of the cutter bar.[45] A steel band 0.51 mm thick with abrasive grit bonded to it was operated at speeds of 8 to 15 m/sec while moving forward at speeds of 0.9 to 1.8 m/sec.[46] Performance with soybeans was best at 15 m/sec and forward speeds above 1.35 m/sec, but satisfactory cutting was not always obtained, as plant severance was incomplete due to insufficient normal force between plant and blade to accomplish sufficiently rapid cutting. When a toothed band saw blade of the same thickness was used at speeds up to 30 m/sec, cutting was satisfactory, and stalk and stubble losses were less than those for a conventional cutterbar. Total power to drive the cutting mechanism amounted to approximately 39 kWhr·cm/t d.m.
5. *Chain-Mowers*. Mowers with cutting blades attached to a continuously moving chain have been proposed.[47,48] Some designs have had opposed sets of blades moving simultaneously in opposite directions. Problems encountered in field tests have been associated with holding minimum clearance between cutting elements and with getting forage to flow over the nearly vibrationless cutterbar.

6. *Rotary or Helical Cutters.* Mechanisms of this type have two main forms, rotating helixes operating against a semicircular guard,[49,50] and that using an elipse mounted at an angle on a rotating shaft so the axial projected view is a circle that matches the full-circular or semicircular guards.[23,51] Energy requirements for the helical device were 9.95 J to cut a soybean stem 0.87 cm in diameter (4.95 kWhr·cm/t d.m.). Rotor torque (4.44 cm radius) to cut the above stem was 3.05 N·m. Field test experiences with this type of device have uncovered problems with the feeding of stems into the cutter, double cutting, ragged cutting, wrapping of forage on the rotating shaft, and acceleration of cut material. Satisfactory field performances has also been achieved under some conditions.

7. *Opposed Rotating Disc Cutter.* Devices of this type have been used for topping sugar beets, for cutting tomato vines for harvest, for cutting tobacco stalks,[52] and for harvesting dry beans with minimum shatter.[53,54] Frequently, a notched disc operates against a smooth edged disc to improve feeding. In some cases, one disc operates opposite a pair of discs to effect double shear. Disc peripheral speed, while usually at or slightly above forward speed, may be as high as 20 times the forward speed. Pairs of rotating square blades have also been used.[29] When these square blades were used to cut corn stalks, forces of about 400 N were found at 12 cm above the ground. Average forces of about 490 N (oriented to pushing the disc axes apart) were found when cutting 2.5-cm diameter tobacco stalks with discs (smooth opposing notched).[52]

8. *High-Speed Disc-Type Impact Cutter.* A study involving a horizontal disc 1 m in diameter with tooth-like notches cut in the periphery was carried out in such a way that soybean stems were caught by the near-radial edges of the notch at peripheral speeds ranging from 36 to 56 m/sec. Satisfactory cutting was achieved and bean shatter losses were about the same as for a conventional floating cutterbar. Shatter was greater with a notch pitch of 152 mm than was the case for pitches of 38 and 76 mm.[55]

9. *Individual Blades Penetrating Rigid Material.* Generally the rigid material is supported by its surroundings, but the force necessary to oppose the penetrating blade is developed and transmitted through the structural strength of the material itself. Blade force is usually expressed in proportion to the width of material cut. The depth of cut, inclination of the tool, and bevel angle of the tool affect the cutting force, which tends to remain constant or vary cyclically during the cut. Prominent among this type of device are the axe and the chisel. Cutting force data were cited for kiln-dry sugar pine (cut depth = 0.76 mm, blade inclination = 60°C to direction of motion) of 273 N/cm for a cross-grain cut, and 75 N/cm parallel to the grain.[2] A cross-grain cut of saturated sugar pine required a force of 161 N/cm. It is believed that saturation helped support concentrated compressive stresses in the zone of the blade edge for the case of the cross-grain cut. These force values represent a range of $0.53 \frac{\text{kWhr·cm}}{\text{t d.m.}}$ (@75 N/cm) to $1.94 \frac{\text{kWhr·cm}}{\text{t d.m.}}$ (@273 N/cm), which values are little different from those found for cutting forage materials. Sugar beets tend not to have the same degree of fiber orientation that is common to wood and forages. Studies of sugar beet topping cutters made using blades with various bevel angles and also using wires of varying diameters were conducted.[56] For bevel angles up to approximately 30°, average forces per unit blade width were about 50 N/cm, and gradually increased with bevel angle thereafter. Forces per unit blade length with wire used as a cutter were 50 N/cm for a wire of 1.6 mm diameter, and this value appeared to change in direct proportion to wire diameter. This value amounts to about $0.604 \frac{\text{kWhr·cm}}{\text{t d.m.}}$.

10. *Dicing Machines.* A combination of blade penetration and chopping is used in machines

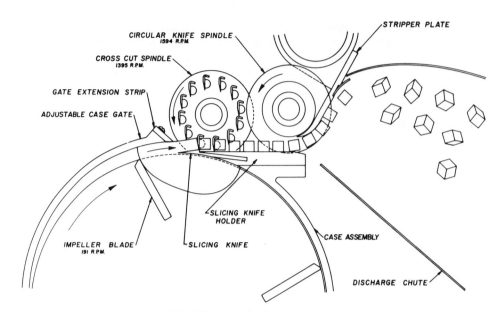

FIGURE 2.   Dicer for food materials. (Taken from Urschel Co., How to Cut Food Products, Urschel Laboratories, Inc., Valparaiso, Ind., 1977. With permission.)

to dice vegetables and other food items (Figure 2). Garden beets could be diced with $0.57 \dfrac{\text{kWhr·cm}}{\text{t d.m.}}$ using a commercial dicer.[57]

11. *Slicing*. In cases where forces in the main direction of blade passage through a material are not sufficient to effect cutting, material failure can be obtained by moving the blade in the direction parallel to its edge line while it is being forced against the material. This causes localized material failure through a microscopic tearing effect. It also reduces the effective bevel angle of the cutting edge. If the angle between the perpendicular to the blade edge and the direction of blade motion relative to the material to be cut is called N (see Figure 1), the effective bevel angle, B_1, is related[5] to the original bevel angle, B, by:

$$B_1 = \tan^{-1}(\tan B \cos N)$$

Furthermore the effective width of the cutting edge, δ', is related[5] to the original cutting edge width, δ, by:

$$\delta' = \delta \cos N$$

Both these factors are responsible for reduction in the forces normal to the blade edge required to achieve cutting when N increases. The coefficients of blade sliding across the material can be designated[5] as ϵ,

$$\epsilon = \tan N = \frac{\text{relative motion parallel to blade edge}}{\text{relative motion normal to blade edge}}$$

The forces normal to and parallel to the blade edge, in terms of a percentage of the original force normal to the blade edge to achieve cutting when N = 0.0, are given below along with relative energy values per unit cross-sectional area of the material cut for varying values of ϵ.[5]

		ϵ = tan N							
	Source	**0.0**	**0.4**	**0.8**	**1.2**	**1.6**	**2.0**	**2.4**	**2.8**
	A	100	98	85	70	54	53	48	43
	B	100	92	73	54	55	50	45	40
	C	100	85	55	45	41	38	35	30
$\frac{\text{Normal force}}{\text{Normal force (N = 0.0)}}$ %	C	100	75	52	38	28	23	22	21
	C	100	99	93	88	76	68	60	56
	C	100	92	82	73	62	54	48	44
	D	100	96	84	80	77	61	54	53
Tangential Force/Normal Force	D	0.0	0.42	0.80	0.80	0.85	0.95	0.87	0.95
Energy/Energy (N = 0.0) %	D	100	101	123	141	156	150	178	172

Slicing experiments were done with forage crops of 15% moisture (w.b.) using normal forces per unit blade length ranging from 8.8 to 18.8 N/cm.[1] Average values for a coefficient of blade friction were 0.805 for a smooth blade and 1.11 for a serrated blade. Energy values for alfalfa were 4.59 $\frac{\text{kWhr·cm}}{\text{t d.m.}}$ with the smooth blade and 3.61 $\frac{\text{kWhr·cm}}{\text{t d.m.}}$ with the serrated blade. Energy values decreased as the normal force increased. These energy values were approximately twice those obtained with conventional shear-bar forage chopper or mower mechanisms. In another series of slicing tests,[11] it was found that energy values were approximately twice those for two-element chopping and ranged from 2.47 to 3.53 $\frac{\text{kWhr·cm}}{\text{t d.m.}}$.

12. *Sawing.* Similar to slicing is sawing, the difference being that the saw makes two cuts — one on either side of the blade — to clear a path for the subsequent passage of the blade structure. The following equation has been given for sawing energy per unit kerf volume.[2]

$$E_s = \frac{(0.377 \times 10^{-6})}{12}\left[B + \frac{A}{t_a}\right]$$

in which E_s = kWhr per cubic inch of kerf, A = blade cutting force per inch of width at zero chip thickness, B = increase in (blade cutting force per inch of width) per unit increase in chip thickness, t_a = chip thickness (inches), and values given for cross-cutting kiln-dried sugar pine were: A = 7 lb/in., B = 971 (lb/in.)/in. For a chip thickness of 0.03 in. or 0.76 mm $E_s = 37.8 \times 10^{-6}$ kWhr/in.^{3} or 2.3×10^{-6} kWhr/cm^3. For a kerf width of 2.5 mm, the energy value amounts to 1.47 $\frac{\text{kWhr·cm}}{\text{t d.m.}}$ which is approximately equal to that for cross-cutting with a penetrating blade. In other tests, hacksaw blades (approximate kerf width = 2.5 mm) with 5.5 and 12.6 teeth per centimeter were used to cut various forage plants at 15% moisture.[1] Energy values for alfalfa were 2.24 and 2.08 kWhr·cm/t d.m., respectively. For all materials cut, the average force parallel to the blade edge was, respectively, for the twotooth-pitch values, 1.93 and 1.43 times the normal force between the blade and the material being cut (normal forces were the same as those mentioned under slicing). Sawing thus appears to consume somewhat more energy than does cutting by mower or shear-bar type devices.

13. *Water-Jet Cutting.* Tests were reported on the use of a water jet to cut lettuce stems

in the field.[58] The pressure used was 41.3×10^6 N/m² and nozzle orifice diameters used were from 0.61 mm to 1.07 mm. When the nozzle was placed within 3.2 cm of the plant and intermittent water spray used, energy consumption of the system amounted to approximately $70 \frac{\text{kWhr·cm}}{\text{t d.m.}}$.

The amount of water used per hectare would range from 122 to 341 ℓ for the most favorable nozzle. The cross-sectional area of lettuce stems to be cut would be approximately 450,355 cm²/ha.

High pressure water jets were also used for cutting sugar-cane stalks.[59] Fully effective cutting occurred only under the following conditions:

1. Pressure 400×10^6 N/m²
2. Nozzle diameter 0.3556 mm or greater
3. Traverse speed 3.2 km/hr or slower
4. Standoff distance 8 cm or less
5. Kinetic energy rate of fluid at nozzle 40.67 kW or greater
6. Fluid flow rate 98.3 cm³/sec

The stalks cut were typically 2.54 cm in diameter, 28% d.m. (approximately half of which was sugar) and had a wet density of 1.103 g/cm³. Based on these figures the cutting energy was 450 kWhr·cm/t d.m.

Water-jet cutting thus appears to have much higher energy requirements than does conventional cutting with blades.

14. *Laboratory Mill.* A common laboratory size-reduction apparatus is a Wiley knife mill, which has rotating radial knives, the axially-oriented edges of which nearly touch those of radially mounted stationary knives. In tests in which third-cutting alfalfa hay was ground in this mill,[20] typical cutting energy values from 0.673 to 1.55 kWhr·cm/ t d.m. were found. When a conventional cylinder-type shear-bar cutterhead was used to cut similar material, values from 1.29 to 1.92 kWhr·cm/t d.m. were measured. The Wiley knife mill using only one stationary knife was less efficient than when six stationary knives were used (2.33 as opposed to 1.40 kWhr·cm/t d.m.).

CUTTING PROCESS VARIABLES

Various researchers have investigated the effects of assorted process variables in the cutting of biological materials. References are categorized in Table 4 according to the variables examined and the materials used in tests of each variable.

Blade Speed

The most commonly investigated variable was blade speed. Generally, with devices that cut between two elements, there was either no change in cutting energy requirements or a very slight increase as blade speed was increased (on the order of 10 to 15% with a doubling of blade speed). Cutting forces tended to show slight decreases as blade speed increased. Flail-type cutting devices require some minimum blade speed for effective cutting. Although some researchers obtained satisfactory cutting at lower speeds, 25 m/sec appeared to be a blade velocity with which most free-standing forage material could be cut without difficulty. Belt and band-type cutters required speeds in the 15 to 40 m/sec range for satisfactory cutting. Flail-type devices tended to display a slight reduction in cutting energy requirements as blade speed increased. Forces and cut material deflections tended to remain unchanged, although in some cases, cutting forces decreased as blade speed increased from that at which unsatisfactory cutting was achieved to more effective cutting velocities. Plant seed shatter

Table 4
REFERENCE CATEGORIZATION BY MATERIAL AND CUTTING PROCESS VARIABLE

Cutting process variable	Material cut			
	Alfalfa	Beans (edible)	Bluegrass	Canary grass
Blade speed	15, 17, 20, 29, 35, 36, 51, 68	53		
Blade bevel angle	13, 14, 40			
Blade dullness	13, 14, 29		24	
Moisture content	3, 4, 7, 8, 10, 12, 14, 15, 17, 27			
Depth of material	9, 10, 14, 15, 17, 29, 61, 66, 67			
Angle between edges	17, 65			
Position along stem	3, 4, 23			
Plant size	9, 23, 29, 35, 66		9, 23	9, 23
Forward speed	26, 40			
Theoretical cut length	7, 8, 9, 24, 67			
Shear-bar profile	10, 29, 65, 66			
Clearance between edges	10, 13, 14			
Blade wear, hardness				
Cutting height	36	53		
Blade serrations	65			
Edge-direction angle	36			
Rake angle	13, 14			
Blade heel clearance	10			
Stalk tilt	66			
Edge path radius	17			
Recutter screen	8, 20			
Fiber grain Orientation				
Material density	3, 12			
Material precompression	1			
Shear-bar orientation				
Crop maturity/generation	4, 9, 20, 23, 36			
Stalk lean	66			
Other	12, 34, 36, 40, 44			

Cutting process variable	Material cut							
	Clover	Corn	Flax	Foxtail	Hay (mixed)	Hemp	Oats	Orchard grass
Blade speed		5, 18, 19, 27, 42, 60			9, 25		29, 37	
Blade bevel angle		5, 13, 19						
Blade dullness		5, 13				5	5	
Moisture content		4, 13						
Depth of material		5		61				
Angle between edges		5, 18					65	
Position along stem		4, 29						
Plant size		27, 60, 66						9, 23
Forward speed	26	26, 27, 42						
Theoretical cut length		8, 9						
Shear-bar profile							65	
Clearance between edges		5, 13, 27					65	
Blade wear, hardness		5	5		5	5	5	

Table 4 (continued)
REFERENCE CATEGORIZATION BY MATERIAL AND CUTTING PROCESS VARIABLE

	Material cut						
Cutting process variable		**Alfalfa**		**Beans (edible)**	**Bluegrass**		**Canary grass**
Cutting height							
Blade serrations					65		9, 32
Edge-direction angle		19, 60					
Rake angle		5, 13					
Blade heel clearance							
Stalk tilt		19					
Edge path radius		18					
Recutter screen		8, 16					
Fiber grain orientation							
Material density		18		12			12
Material precompression			1				
Shear-bar orientation		18					
Crop maturity/generation		4, 42					9, 23
Stalk lean		29, 66					
Other	41	5	5	5, 12, 24	5	5	12

	Material cut							
Cutting process variable	**Quack grass**	**Ragweed**	**Rape**	**Rice**	**Rye**	**Rye grass**	**Sod grasses**	**Sorghum**
Blade speed			35	30			37, 38, 58	
Blade bevel angle				30				
Blade dullness	9, 32				69			
Moisture content				28		31	31, 58	
Depth of material				28				
Angle between edges	9, 32					65	58	
Position along stem				30				
Plant size		66				31	31, 58	
Forward speed						38	38	
Theoretical cut length								
Shear-bar profile						65		
Clearance between edges						65		
Blade wear, hardness					5, 69		5, 69	
Cutting height								
Blade serrations	9, 32			28		65		
Edge-direction angle							58	
Rake angle								
Blade heel clearance								
Stalk tilt								
Edge path radius								
Recutter screen								
Fiber grain orientation								
Material density								
Material precompression								
Shear-bar orientation								
Crop maturity/generation						31	31	
Stalk lean		66						
Other				28, 44	5, 69		5	8, 50

Table 4 (continued)
REFERENCE CATEGORIZATION BY MATERIAL AND CUTTING PROCESS VARIABLE

	Material cut				
Cutting process variable	**Soybeans**	**Sudan grass**	**Sugar beet**	**Sugar cane**	**Sunflower**
Blade speed	42, 46, 49, 54, 55				33—35
Blade bevel angle			56		
Blade dullness			5, 56		5, 34, 35
Moisture content	49				
Depth of material					
Angle between edges	49				
Position along stem					
Plant size	49				35
Forward speed	42, 46, 49	46	56	59	
Theoretical cut length					
Shear-bar profile					
Clearance between edges					
Blade wear, hardness					5
Cutting height	49				35
Blade serrations					
Edge-direction angle		36			
Rake angle					
Blade heel clearance					
Stalk tilt					
Edge path radius					
Recutter screen					
Fiber grain orientation					
Material density					
Material precompression					
Shear-bar orientation					
Crop maturity/generation					
Stalk lean					
Other		36, 40		43, 59	5

	Material cut					
Cutting process variable	**Timothy**	**Tobacco**	**Wheat**	**Wood**	**Wool**	**General**
Blade speed	1, 61	52	35	39	5	5
Blade bevel angle	1, 11, 30, 61		5		5	5, 6, 62, 63
Blade dullness	1, 11, 61		40	39		6
Moisture content	1, 61			2		
Depth of material	1, 61			39		5
Angle between edges	1, 11, 61		9, 32			5, 6
Position along stem						
Plant size	1, 9, 30, 66	52				
Forward speed		52				32
Theoretical cut length	11			2, 39		
Shear-bar profile	1, 11, 29, 66					
Clearance between edges	1, 11, 61					5
Blade wear, hardness			5, 69	5		5, 6, 9, 62, 64
Cutting height	1					
Blade serrations	1, 61		9, 32, 69			
Edge-direction angle					5	5, 32
Rake angle	11			2		5

Table 4 (continued)
REFERENCE CATEGORIZATION BY MATERIAL AND CUTTING PROCESS VARIABLE

		Material cut		
Cutting process variable	Alfalfa	Beans (edible)	Bluegrass	Canary grass
Blade heel clearance			2	
Stalk tilt	66			
Edge path radius				
Recutter screen				
Fiber grain orientation			2, 39	
Material density				
Material precompression	1			
Shear-bar orientation				
Crop maturity/generation				
Stalk lean	66			
Other		5, 50	5	

either decreased or remained unchanged with blade speed increases for impact cutting devices. The peak torque requirements on helical or elliptical cutting devices decreased as rotational speed increased, though this may have been influenced partly by instrumentation characteristics. Generally, cutting force and energy values for all types of devices that cut effectively changed little with blade speeds in the range from quasi-static to 60 m/sec.

Blade Bevel Angle (See Angle B in Figure 1)

For cutting any kind of material, the blade bevel angle found most effective in practice tends to increase with the hardness of the material cut. This determination is mainly based on wear considerations. For common crop materials, angles of 20 to 30° seem best suited. Very little reduction in force and energy values can be expected by using angles less than 20° and wear rates for small angles would be higher. Above 30°, there is a tendency for both force and energy values to increase (70 to 80° blades having values twice that of a 25° blade). This effect is particularly noticeable when cutting greater depths or thicknesses of material (see equation in section on Parameters of Cutting Performance).

Blade Dullness (Edge Radius Increase)

There appears to be distinctly lower cutting force and cutting-energy values associated with very sharp blades. This may be due to these very sharp blades cutting the cells of the material individually without compressing the material noticeably, avoiding simultaneous failures of large groups of cells. Edge widths or diameters of approximately 0.25 mm tend to have approximately double the force and energy requirements of edges with widths in the 0.05 to 0.1 mm range. A general relationship[5] has been expressed in which the energy required to effect cutting of a given cross-section of material is in proportion to the square root of the edge radius. Edge widths less than 0.02 mm are difficult to maintain while cutting crop materials because the forces required produce stress levels, which can cause failure in steel blades having edge widths less than approximately this value. High velocity impact cutting devices appear to be slightly less sensitive to blade dullness than do dual-element cutting devices. However, the minimum blade velocity required for effective cutting with such devices increases as the blade becomes duller (a 40% minimum velocity increase as blade edge thickness increased from 0.25 to 1.0 mm). The increase in edge thickness from 0.25 to 1.0 mm increased cutting forces approximately 25% in tests of a high velocity impact cutter.

Moisture Content

A general tendency exists for cutting forces to increase slightly with major decreases in plant moisture content. As moisture content changed from turgid (60 to 80% w.b.) to air dry (10 to 20% w.b.), cutting force increases of from 20 to 50% were found by various researchers. Similarly, tensile failure stresses for alfalfa stems were found to be greater at lower moisture contents than at high moisture contents. No such change in tensile strength was found for corn stalks. Just the opposite trend was found with regard to cutting energy per unit of d.m. cut to a given length. The range of experience with this relationship was bounded on one side by several cases in which there was no significant change in cutting-energy values per unit d.m. over a range of moisture contents, and on the other side by some cases in which cutting-energy values per unit wet matter did not change with moisture content. A median trend could be approximated by cutting-energy values per unit d.m., which were 50% higher at 65% w.b. than they would be at 30% w.b.

Depth of Material Cut (See Dimension E in Figure 1)

Material depth can be specified for a pile of biological material in terms of the mass of d.m. per unit of area of the pile base surface, i.e., grams per square centimeter — g/cm^2. It is believed that as the depth of material cut increases, stress concentration near the blade is diminished and consequently greater precompression energy is expended in achieving the stress level necessary to accomplish cutting. Example data with silage corn[5] showed that the proportion of total cutting energy used just to compress the material to the point where cutting could begin, increased from 35 to 59% as the depth of material increased from 2.5 to 12.0 cm. This pertains to the cutting process between two elements, in which cutting resistance does not come from structural strength or inertial resistance of the material itself. In cases where cutting forces have been measured, quadrupling material depth approximately doubles maximum cutting forces. Tests in which material depth has been doubled have shown changes in values of cutting energy per unit d.m. (cut to a given length) ranging from a 17% decrease to a 50% increase, with a 20% increase as an approximate median value. Cases in which material depth were quadrupled resulted in 50 to 100% increases in cutting energy values per unit d.m. cut to a given length. Illustrative values of cutting force and energy values are as follows:

| | | Moisture | | Material depth — g (d.m.)/cm² | | | |
| | | Percent | | | | | |
Item	Material	(w.b.)	Units	0.12	0.24	0.36	0.48
Max. force	Timothy	20	N/cm	140	184	219	254
Energy	Timothy	20	kWhr·cm d.m.	1.77	2.12	2.50	2.79
Energy	Alfalfa	20	kWhr·cm d.m.	1.58	2.20	2.54	2.66

Included Angle Between Edges (See Angle F in Figure 1)

In devices like mowers, if this angle is too great, material will be expelled during cutting. Included angles as small as 27 to 34° were necessary to retain free straw between smooth edges (32 to 40° for grass). Having both edges serrated increased this angle to 66 to 84° (straw) and 84 to 90° (grass). Maximum cutting forces are reduced by increasing the included angle (mainly due to distributing cutting work over greater blade motion) on the order of 1% per degree increase. Cutting energy per unit d.m. cut to a given length was in some cases found to increase with increases in included angle (33% increase with a change from 4 to 18°, for example). In other cases, minimum cutting energy values, which were slightly lower than those with zero, included angle occurred with angles in the 25 to 50° range.

Position of Cut Along the Stem

There is evidence that both corn and alfalfa have higher d.m. densities in lower sections of the stem than they do at higher sections. Thus, when cutting force or energy is related to the measured cross section of the material, these values for cutting near the tip of the plant may be only 40% of those for cutting near the base.

Forward Speed of Field Cutting Device

Generally, no increase in values of cutting energy per unit of d.m. cut to a given length were noticed as forward speed increased. In a number of field machines, total power input increased with forward speed, but much less rapidly than forward speed itself. Thus, either cutting-energy values per unit material decreased or a sizable proportion of the measured energy input was not due to the cutting process alone. Tests with a flail-type forage harvester indicated that mean cut-particle length increased as forward speed increased.

Length of Material Cut Off from Main Piece (See Dimension I in Figure 1)

It is frequently assumed that, for the lengths of cut used for crop materials (6 mm or longer), values for cutting force and energy per cut are not affected by length of material cut. This assumption is embodied in the following method for determining cutting energy:[9]

$$\frac{\text{Cutting Energy}}{\text{Mass Cut to Standard Length}} = \frac{W_1 - W_2}{(1/C_1) - (1/C_2)}$$

where W_1 = energy to cut one unit of mass to length C_1, W_2 = energy to cut one unit of mass to length C_2, and C = theoretical length of cut. Power measurements on many machines show that power requirements frequently do not double when cutting lengths are halved, indicating that either shorter lengths are cut with less energy or that measurements include energy delivered to functions not directly affected by additional cutting requirements. Forces to cut wood tend to increase linearly with chip thickness, and wood sawing energy per unit kerf volume less than doubles as chip thickness is halved, indicating that less energy is used in cutting a short length than a long one.

For shear-bar type forage choppers, actual mean length of grass particles tends to average about 1.5 times the theoretical length of cut. For corn stalks, mean particle length tends to be on the order of 1.3 times the theoretical length of cut.

Shear-Bar Profile

A sharpened shear bar with a narrow (20 to 30°C) bevel angle, when working against a dull or square-edged blade, causes cutting-force and energy values of about the same magnitude as when a sharp blade is working against a square shear bar. A sharpened shear bar working against a sharp blade can cause only slight (0 to 25%) reductions in force and energy values as compared to when a square shear bar is used. However, a square shear bar working against a square or dulled (0.28 mm edge radius) blade tends to have both force and energy values approximately twice those when a sharpened shear bar is working against the same sort of blades. As the edge radius of a square shear bar increases, there appears to be only a slight increase in cutting energy (a radius increase from 0.02 mm to 0.6 mm causes a 14% cutting energy increase, for example). It has been found that if hard-facing material is to be applied to help maintain a given shear-bar profile, the best procedure is to apply it to the vertical edge — the face of the bar that the knife edge sweeps across.

Clearance Between Cutting Edges (See Dimension K in Figure 1)

When blades are sharp there seems to have been little effect on cutting force and energy values due to increased clearance between cutting edges when maximum values examined were 0.26, 0.406, 0.625, 0.889, and 2.32 mm. However, when large clearance values were combined with square-edged or dulled blades, cutting-energy values could as much as double. This combination was also found to produce poor-quality cutting with ragged or torn plant fibers. Even with sharp blades, example data with corn show that cutting energy values increased 67% as clearance increased from 2 to 10 mm.

Cutting Height Above Plant Base

Cutting energy for a helical mower decreased 20% as cutting height increased from 2.5 to 6.3 cm. An impact cutting device was much more effective in cutting at a 0.6 cm height than at a 7.6 cm height. A free-standing sunflower stalk cut by a blade moving at 15 m/sec had a minimum cutting force of about 88 N when cut 150 mm above the point of support. When cut at a height of 10 mm, the cutting force was about 108 N. An impact cutter striking stems at increased heights caused increased stem deflection — twice as much at a 10-cm height as the 0.94-cm deflection at a 5-cm height. A dry-bean cutting device incurred a 7.2% crop loss by increasing cutting height from zero to 5 cm.

Blade Serrations

Generally, cutting force and energy values for under-serrated blades tend to be from 35 to 75% higher than for smooth blades in similar conditions. The reason is that at the root of the serration groove, the cutting plane has more clearance than at the outer blade edge, and thus cutting tends to take place over a zigzag line through the material. The presence of serrations does, however, allow approximately twice as large included angles between cutting edges without expulsion of stalks, in comparison with two smooth-edged blades or a smooth blade and a serrated ledger plate. Under-serrated blades are slightly more effective in this respect than top-serrated blades, and under-serrated blades can retain their serrations after sharpening. In a comparison between smooth and serrated sickles for cutting rice straw, there were, on the average, no differences in cutting force or cutting-energy values. However, cutting energy requirements with the serrated sickle were more sensitive to the size of the group of stalks cut. In cutting hay by slicing, serrated blades required less than half the energy of smooth blades, but the slicing force (with a given normal force) was approximately 40% greater.

Angle of Blade Edge with the Direction of Blade Motion (90° Minus Angle N in Figure 1)

When growing plants are deflected by a moving edge, the stem will not slide along the edge (i.e., it will follow the direction of motion of the edge) if the direction of blade motion does not deviate from the perpendicular to the blade edge by an amount greater than the arc tangent of the (coefficient of friction of plant on edge). Once the direction of blade motion deviates from this perpendicular by an angle greater than that specified above, the plant stems will assume a horizontal line of deflection that is aligned with the force applied by the blade. This force will then make an angle of the arc tangent of the (coefficient of friction) to the perpendicular to the blade edge irrespective of the direction of blade motion. When cutting standing stalks with flail-type impact devices at intermediate velocities (2 to 10 m/sec), best results were obtained when the blade edge made an angle of from 30 to 60° with the direction of motion. At angles smaller than 30° incomplete severance of stalks resulted. At angles greater than 60°, cutting force and energy values increased. A discussion of cutting parameters with varying angles between the blade edge and the direction of blade motion appears under the section on cutting by slicing.

Rake Angle (See Angle O in Figure 1)

Rake angle is the angle between the leading face of the blade and the plane perpendicular

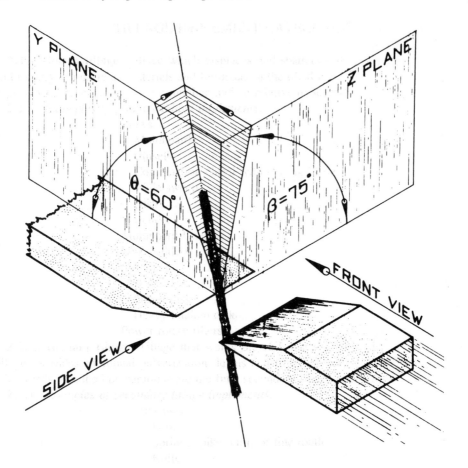

FIGURE 3.   Illustration of angles which a stem may assume between cutting blades. As defined in the text: "tilt" = 90° − β and "lean" = 90° − θ. (From Prince, R. P. and Wheeler, W. C., Some Factors Affecting the Cutting of Forage Crops, Prog, Rep. No. 37, Agricultural Experiment Station, University of Connecticut, Storrs, 1959. With permission.)

to the direction of blade movement. Of the three studies reviewed concerning this factor in shear-bar type machines, one showed almost no effect of rake angles between 32 and 67° on cutting energy values, while a second indicated that an increase in rake angle from 45 to 60° resulted in a 20% reduction in cutting forces and a 30% reduction in cutting energy values. The third showed that when cutting corn silage, increases in rake angle from 30 to 70° resulted in a 70% reduction in cutting energy per unit cross-section of material. For tools cutting wood, an increase in rake angle from 5 to 30° generally resulted in an approximate halving of forces on the tool in the direction parallel to tool motion.

Stalk Tilt

If a stalk lies in the plane perpendicular to the direction of blade motion, the extent to which the stalk is rotated in that plane away from the perpendicular to the blade edge, can be defined as the stalk tilt (see Figure 3). Data on corn, alfalfa, and timothy indicate that cutting-energy values are reduced by increasing stalk tilt values from zero. A 15° tilt angle for timothy and alfalfa reduced cutting energy about 15%. With corn, increasing tilt angle form 10 to 40° reduced the energy to sever the stalk by 14% and the maximum force by 22%. Further increases in tilt angle to 50° increased requirements for cutting force and energy.

Plant Size

There is limited evidence that cutting forces per unit cross-sectional area increase slightly with increases in plant stem diameter (19% with a doubling of diameter in one case). Tests with a free-standing sunflower stalk cut by a blade moving at 16.5 m/sec indicate an approximately linear relationship between cutting force and stem diameter (about 12 N force increase per millimeter of stem diameter increase). Cutting energy per unit d.m. cut to a given length also tends to increase with stalk diameter (42% with a doubling of diameter in one case). This phenomenon may be due to the fact that larger stems may be more mature and have a higher d.m. density than smaller-diameter stems.

Crop Maturity or Generation

Tests with third-cutting alfalfa and second-cutting alfalfa indicated that the cutting energy, as determined with a Wiley mill was approximately 71% greater for the second-crop material than for the third-crop material, on the basis of a given amount of d.m. cut to a given length. Third-crop material is characterized by finer texture and thinner stems than second-crop material. Energy to cut ryegrass stems of a given cross-sectional area increased about 15% with each week of additional age at approximate maturity.

Edge Path Radius (See Dimension S in Figure 1)

Tests with cylinder-type shear-bar cutterheads of varying radii (23 to 36 cm) indicated both slight increases and slight decreases in cutting-energy values with radius increase. The data are sufficient, however, to indicate that little change in cutting energy values can be expected from radius changes of this magnitude.

Recutter Screens

In some cases, cylinder-type cutterheads are fitted with recutter screens having either round holes or diagonally-oriented slots to further reduce cut-particle length. Data indicate that when cutting energy is computed per unit d.m. cut to a given length, there are a few differences in such a parameter when machines with and without recutter screens are compared. If the recutter screens are relied upon for an approximate halving of mean particle length, about a 10% increase in cutting energy values can be expected.

Fiber Grain Orientation

Fundamental cutting measurements with wood indicate that blade forces (per unit blade width) when cutting across the grain are approximately two to three times those found when cutting parallel to the grain.

Material Density

The ultimate tensile strength of alfalfa stems appeared in laboratory tests to be directly proportional to the d.m. bulk density of the material. Similarly, the energy to shear silage materials from alfalfa, timothy, and orchard grass appeared to be directly proportional to the amount of d.m. on the cross section to be sheared (shearing energy per unit area was directly proportional to d.m. bulk density). This, however, was not the case with regard to maximum shearing force (for a given d.m. cross section). Shearing force increased with increases in d.m. density, but at a rate which would indicate a shearing strength of about 1.2×10^6 N/m^2 with a d.m. density of zero.

Precompression of Material

A portion of the energy supplied in cutting goes to compress the material prior to cutting. The cutting of precompressed material thus exhibits lower cutting-energy requirements than for the same material without precompression (51% reduction for alfalfa, 32% for timothy).

Cutting forces with a zero-degree included angle between blade edges were not reduced by precompression, but those with a 25° angle were reduced by 20% for both materials due to the blade being in contact with less material at any given moment.

Shear-Bar Orientation (See Dimension X in Figure 1)
Normally the shear bar of a cylinder-type cutterhead is placed so that a plane parallel to its surface passes through the center of the cylinder. Tests with this plane passing 2.5 cm above, 2.5 cm below, as well as directly through the cylinder centerline were conducted. Cutting-energy values were about 20% higher when the plane was arranged to pass below the cylinder centerline than with the other two orientations.

Stalk Lean
If a plane is constructed parallel to the direction of blade motion and perpendicular to the blade edge, the amount of rotation of a stalk in that plane from a position that is perpendicular to the direction of motion may be defined as stalk lean (see Figure 3). Laboratory tests indicate that the energy to cut a stalk of a given size may be reduced slightly (0 to 15%) for a lean of 30° as compared with that of 0°, with timothy and alfalfa. A lean of 45° can reduce cutting energy values for corn by about 20% and those for ragweed by about 25% as compared to the zero-lean case.

Back of Blade Clearance or Relief (See Angle Z in Figure 1)
In most cutting processes in which the positions of the tool and of the material can be controlled relative to each other, it is considered desirable to have clearance between the tool and the cut material behind the cutting edge. Tests with tools for cutting wood show that a clearance angle of 15°C can reduce the tool force in the direction of motion about 10% from that with zero clearance angle (rake angle held constant). Clearance angles greater than 25°C degrees can increase tool force above that found with 0°C. It has been found that built-in sharpeners in cylinder-type cutterhead machines tend to produce negative clearance angles if the cylinder is turned in the normal direction while the sharpener is used. If the cylinder is turned in the reverse direction while sharpening, a small, positive clearance angle is produced. Shear bars also can be constructed with a clearance angle (see angle J in Figure 1). Tests using a shear bar with a 20° clearance angle had no effect on cutting-energy values, but did cause problems with forage wrapping in an uncut state over the shear-bar edge when blade-to-bar clearance became large.

Blade Hardness and Wear
The typical analysis of steel used as blade material for a mower knife is as follows: C — 0.82%, Si — 0.13%, Mn — 0.51%, P — 0.016%, S — 0.016%, Cr — 0.12%. Material hardness is approximately 55 to 59 on the Rockwell C scale for the edges. The blades are usually hardened in the edge area with a nitriding process leaving the blade center with an approximate hardness of 25 to 35 on the Rockwell C scale. Forage chopper cutterhead knives are frequently made of an alloy steel and hardened in the edge area to 50 to 54 on the Rockwell C scale.

Experiments with a mower working in a rye field showed that knife-blade edges wore back (linearly with area cut) to approximately 0.30 mm from their original dimensions after 41.1 hr of work, during which the mower advanced 142 km. In this worn state, the 19°C bevel-angle blade had an edge radius of approximately 0.03 mm. Wear toward the forward tip of the blade was much more pronounced than that toward the rear part of the blade. Blades having an edge hardness of 63 (Rockwell C) experienced only about 60% of the wear of blades having an edge hardness of 55 (Rockwell C). Tests based on the energy required to cut a standard piece of paper showed that mower blades with bevel angles of

14 to 20°, when used for 4 hr of hay cutting, required approximately 4% more energy to cut the paper than when the blades were new. In another test with flywheel-type forage chopper knife blades, cutting-energy requirements increased 24% in 15 hr for a 14° bevel-angle blade as compared with that for a 24° bevel-angle blade.

Data on the rate of wear for blades cutting various materials[5] are compared below, with similar blade wear data for corn chopped for silage taken as a standard of 1.0.

Material	Relative wear	Material	Relative wear
Sunflower	1.16	Wheat straw	1.84
Winter wheat	1.08	Hay	1.62
Rye	0.79	Grass	0.65
Flax	1.05	Vetch-oats	0.70
Hemp	1.04	Wood	7.35

Thickness of Blade (See Dimension BB in Figure 1)

When cutting thin layers of material, blade thickness is not expected to have much effect. However, when cutting very thick layers of material, a thicker blade will require greater displacement of the material to allow passage of the blade. Example data when cutting a 10-cm thick layer of corn for silage indicated a 46% increase in energy requirements per unit cross-sectional area as blade thickness increased from 2 to 8 mm.

CUTTING CHARACTERISTICS OF VARIOUS MATERIALS

Information in Tables 1, 2, and 3 indicate that force and energy parameters to cut common biological materials tend to fall within fairly constrained limits. In many cases, these parameters are mainly influenced by the characteristics of the mechanism with which the measurements were made. References are categorized in Table 4 according to the various materials cut in tests, and according to the cutting process variables associated with each material.

Alfalfa

Data for cutting energy and force values for alfalfa appear in Tables 1, 2, and 3. Of those values that are believed to be free from complicating factors, the mean value for cutting energy is $1.58 \frac{\text{kWhr·cm}}{\text{t d.m.}}$. The range was from 0.17 to $4.06 \frac{\text{kWhr·cm}}{\text{t d.m.}}$. The average cutting force was $639 \frac{\text{N·cm}}{\text{g d.m.}}$ (range 338 to $1257 \frac{\text{N·cm}}{\text{g d.m.}}$).

Bluegrass

The energy required to cut a bluegrass stem was found to be approximately twice that needed to cut an alfalfa stem of the same diameter.

Canary Grass

The energy required to cut a stem of Reed canary grass was found to be approximately 23% greater than that to cut an alfalfa stem of the same diameter.

Clover

The energy to cut ladino clover with a conventional mower was 1.32 times that for alfalfa and 0.6 times that to cut corn for a given unit of d.m. cut to a given length. Energy to cut red clover with a flail-type forage harvester was 0.75 times that for alfalfa. Energy required to cut a stem of ladino clover was 40% of that to cut an alfalfa stem of the same diameter.

Corn

Data for cutting energy and force values for corn appear in Tables 1, 2, and 3. Of those values that are believed to be free from complicating factors, the mean value for cutting energy is $1.43 \frac{\text{kWhr·cm}}{\text{t d.m.}}$. The range was from 0.58 to $2.70 \frac{\text{kWhr·cm}}{\text{t d.m.}}$. The average cutting force was $174 \frac{\text{N·cm}}{\text{g d.m.}}$ (range, 70 to $248 \frac{\text{N·cm}}{\text{g d.m.}}$). The force to cut a single corn stalk averaged 191 N (range 49 to 312 N). The force to cut a layer of silage corn stalks with an unbeveled blade having a 0.10 mm edge thickness, ranged from 79.4 to 91.2 N/cm of blade length (mean, 85.3 N/cm).

Flax

The force per unit blade length to cut flax was 1.3 times that for cutting silage corn.

Foxtail

The energy to cut foxtail was approximately 68% of that for alfalfa, while the cutting force for foxtail was 1.09 times that for alfalfa — all on a d.m. basis.

Fruit (Various)

Cutting energy to dice the following fruits[57] is expressed in $\frac{\text{kWhr·cm}}{\text{t d.m.}}$ and is based on the surface area of the cubes divided by two to get the cross-sectional area cleaved.

$$\text{Apples} \quad (84.9\% \text{ moisture, w.b.}) = 0.466 \frac{\text{kWhr·cm}}{\text{t d.m.}}$$

$$\text{Canteloupe} \quad (92.8\% \text{ moisture, w.b.}) = 0.977 \frac{\text{kWhr·cm}}{\text{t d.m.}}$$

$$\text{Pineapple} \quad (85.3\% \text{ moisture, w.b.}) = 0.478 \frac{\text{kWhr·cm}}{\text{t d.m.}}$$

Hemp

The force per unit blade length to cut hemp was 1.34 times that for cutting silage corn.

Meats (Various)

Cutting energy to dice the following meats[57] is as follows:

$$\text{Beef} \quad (76.3\% \text{ moisture, w.b.}) = \frac{\text{kWhr·cm}}{\text{t d.m.}}$$

$$\text{Chicken} \quad (70.14\% \text{ moisture, w.b.}) = 1.38 \frac{\text{kWhr·cm}}{\text{t d.m.}}$$

$$\text{Codfish (frozen)} \quad (82.6\% \text{ moisture, w.b.}) = 2.38 \frac{\text{kWhr·cm}}{\text{t d.m.}}$$

$$\text{Tuna} \quad (72.7\% \text{ moisture, w.b.}) = 1.02 \frac{\text{kWhr·cm}}{\text{t d.m.}}$$

Oats

The only data available on cutting oats are those obtained with impact cutting devices. Oats in the dough stage required 46% of the cutting energy per unit of stem cross-sectional area as did alfalfa. Oat straw required 88% as much cutting energy as did sod grasses on a d.m. basis. The force per unit blade length to cut an oats-vetch mixture was 0.42 times that for cutting silage corn.

Orchard Grass

The energy required to cut a stem of orchard grass was found to be approximately 67% of that required to cut an alfalfa stem of the same diameter. Cutting energy per unit d.m. cut to a given length, in another test, was 12% greater for orchard grass than for alfalfa, while the cutting force for orchard grass, per unit of d.m. on the cross section was 6% higher than that for alfalfa.

Ragweed

The energy to cut a ragweed stem of a given diameter was 1.5 times that to cut an alfalfa stem of the same diameter. Cutting energy on an equivalent diameter basis for ragweed was 1.7 times that for corn.

Rape

The force to cut a free-standing rape stalk using a high velocity blade was 13% lower than that for an alfalfa stalk per unit cross-sectional dry matter.

Rice Straw

Two independent tests cutting moist rice straw with mower blades had an average cutting-energy value of $1.47 \frac{kWhr \cdot cm}{t \ d.m.}$ and an average cutting force of $610 \frac{N \cdot cm}{g \ d.m.}$.

Rye

The force per unit blade length to cut rye was 0.79 times that to cut silage corn.

Ryegrass (Perennial)

The energy per unit d.m. on a given cross section to cut perennial ryegrass of 75% w.b. was 0.44 kWhr·cm/t d.m..

Sorghum

Energy requirements to chop sorghum for silage (per unit d.m. cut to a given length) were 86% of that for corn.

Soybeans

Energy requirements for chopping green soybean plants with a flail-type forage harvester were 57% of those for corn when adjusted to a basis of energy per unit d.m. cut to a given length. On this same basis, the energy to cut soybeans with band-saw blade operating at 30 m/sec was very nearly the same as required for the flail-type machine operating at from 30 to 49 m/sec, i.e., 39 and 36 kWhr·cm/t d.m., respectively.

Sudan Grass

Energy requirements for impact cutting of Sudan grass stems were 71% of those for alfalfa stems on a per unit cross-sectional area basis. Cutting-energy requirements for a flail-type forage harvester for Sudan grass were 75% of those for alfalfa.

Sugar Beets

A typical cutting force for sugar beets was 50 N/cm width of blade. This is equivalent to a cutting-energy requirement of $0.604 \frac{kWhr \cdot cm}{t \ d.m.}$, based on a beet moisture content of 77%.

Sunflower

Energy requirements to cut one sunflower stem with a high-velocity impact cutter were

$3.71 \dfrac{\text{kWhr·cm}}{\text{t d.m.}}$. The cutting force per unit of dry matter on the stem cross section was 545

$\dfrac{\text{N·cm}}{\text{g d.m.}}$ based on a d.m. bulk density of 0.14 g/cm³. This force value was about 42% greater than for an alfalfa stem. However, on the basis of force per unit stem diameter, the value of 73 N/cm for sunflower was only 54% of that for alfalfa. The force per unit blade length to cut sunflower was 1.09 times that to cut corn for silage.

Timothy

Cutting energy values for timothy averaged $1.26 \dfrac{\text{kWhr·cm}}{\text{t d.m.}}$, and cutting energy values averaged 81% of those for alfalfa in comparison cases. Cutting-force values for timothy averaged $519 \dfrac{\text{N·cm}}{\text{g d.m.}}$, and cutting force values averaged 84% of those for alfalfa.

Tobacco

Cutting forces for tobacco stems were 489 N for a stem diameter of 2.54 cm. If d.m. density of the stem is assumed to be 0.3 gm/cm³, the cutting force above is the equivalent of $322 \dfrac{\text{N·cm}}{\text{g d.m.}}$.

Vegetables (Various)

Cutting energy to dice the following vegetables[57] is based on the surface area of the cubes divided by two to get the cross-sectional area cleaved.

Vegetable	Moisture content (% w.b.)	kWhr·cm / t d.m.
Beets	87.6	0.568
Cabbage	92.4	1.924
Carrots	88.2	0.502
Celery	94.5	4.800
Onions	87.5	0.845
Potatoes	77.9	0.319
Turnips	90.9	0.774

Wheat

Energy requirements to cut one wheat stem with a high velocity impact cutter was 9.2 kWhr·cm/t d.m. The cutting force per unit of d.m. on the stem cross section was 952 N·cm/ g d.m. — more than twice the corresponding figure for alfalfa. However, on the basis of force per unit stem diameter, the value of 57 N/cm was less than half of that for alfalfa. The force per unit blade length to cut fresh winter wheat was 0.72 times that to cut corn for silage, while the force per unit blade length to cut wheat straw was 1.28 times that to cut silage corn.

Wood

Cutting-force and energy values for wood are discussed in detail in the sections devoted to cutting by: individual blades penetrating rigid material, and sawing. The average cutting-energy value of those cited is $1.31 \dfrac{\text{kWhr·cm}}{\text{t d.m.}}$ and a typical cutting force is 170 N/(cm of blade length) as based on a cutting depth of 0.076 cm in kiln-dry sugar pine. One series of tests reports that the force per unit blade length to cut wood ranged from 26 to 687 N/cm when using an unbeveled blade of 0.1-mm edge thickness. The midpoint of this range is 4.18 times the corresponding value for cutting silage corn. Wood is a material for which cutting forces (and consequently cutting energy parameters) increase as the depth of cut is increased. Flail cutters operating in wood and cutting the wood to lengths of approximately 4 cm had cutting-energy requirements on the order of $5.34 \dfrac{\text{kWhr·cm}}{\text{t d.m.}}$.

REFERENCES

1. **Chancellor, W. J.,** Basic Concepts of Cutting Hay, Ph.D. thesis, Cornell University, Ithaca, New York, 1957, 170.
2. **Koch, P.,** *Wood Machining Processes,* Ronald Press, New York, 1964, 530 pp.
3. **Halyk, R. M. and Hurlbut, L. W.,** Tensile and shear strength characteristics of alfalfa stems, *Trans. ASAE,* 11(2), 256, 1968.
4. **Prince, R. P., Bartock, J. W., and Broadway, D. W.,** Shear stress and modulus of elasticity of selected forages, *Trans. ASAE,* 12(4), 426, 1969.
5. **Reznik, N. E.,** Teoriya Rezaniya Lezviem i Osnovy Rascheta Rezsushchikh Apparatov (Theory of Blade Cutting and Basis of the Calculations of Cutting Apparatus) Mashiostroenie, Moskova, 1975. (Translated from Russian for the U.S. Department of Agriculture and National Science Foundation, Washington, D.C. by the Al Ahram Center for Scientific Translation, 1979.)
6. **Stroppel, Th.,** Zur Systematick der Technologie des Schneidens, *Grundlagen der Landtechnik,* 5, 120, 1953.
7. **Blevins, F. Z. and Hansen, H. J.,** Analysis of forage harvester design, *Agric. Eng.,* 37(1), 21, 1956.
8. **Hennen, J. J.,** Power Requirements for Forage Chopping, ASAE Paper No. 71-145, American Society of Agricultural Engineers, St. Joseph, Mich., 1971.
9. **Richey, C. B., Jacobson, P., and Hall, C. W.,** *Agricultural Engineers' Handbook,* McGraw-Hill, New York, 1961.
10. **Liljedahl, J. B., Jackson, G. L., De Graff, R. P., and Schroeder, M. E.,** Measurement of shearing energy, *Agric. Eng.,* 42(6), 298, 1961.
11. **Berentsen, O. J.,** Energy requirements for grass chopping, Res. Rep. No. 22, 1432 Ås-NLH, *Norwegian Inst. Agric. Eng.,* 52(28), 1, 1973.
12. **Bright, R. E. and Kleis, R. W.,** Mass shear strength of haylage, *Trans. ASAE,* 7(2), 100, 1964.
13. **Ige, M. T. and Finner, M. F.,** Optimization of the performance of the cylinder-type forage harvester cutterhead, *Trans. ASAE,* 19(3), 455, 1976.
14. **Ige, M. T. and Finner, M. F.,** Effects and interactions between factors affecting the shearing characteristics of forage harvesters, *Trans. ASAE,* 18(6), 1011, 1975.
15. **Moustafa, S. M. A., Searey, S. W., and Brusewitz, G. H.,** Development of a forage chopping energy standard procedure, *Proc. 1st Int. Grain Forage Handling Conf.,* American Society of Agricultural Engineers, St. Joseph, Mich., 1978, 261.
16. **Dernedde, W. and Peters, H.,** Wirkung und Leistungsbedarf von Nachschneidensystemen fur Exakthacksler, *Grundlagen Landtechnik,* 26(1), 23, 1976.
17. **Springer, A. G., Smith, J. L., and Tribelhorn, R. E.,** Forage Harvester Cutterhead Kinetics, ASAE Paper No. 76-1008, American Society of Agricultural Engineers, St. Joseph, Mich., 1976.
18. **Tribelhorn, R. E. and Smith, J. L.,** Chopping energy of a forage harvester, *Trans. ASAE,* 18(3), 423, 1975.
19. **Prasad, J. and Gupta, C. P.,** Mechanical properties of maize stalk as related to harvesting, *J. Agric. Eng. Res.,* 20(1), 79, 1975.
20. **Von Bargen, K., Verma, L. R., and Foster, W. E.,** Particle Size Reduction of Alfalfa Hay by Cutting, ASAE Paper No. 79-1541, American Society of Agricultural Engineers, St. Joseph, Mich., 1979.
21. **Hochstein, R. R., Thauberger, J. C. and Nyborg, E. O.,** Report on New Holland 840 Forage Harvester, Bull. E0378A, Prairie Agricultural Machinery Institute, Portage la Prairie, Manitoba, 1979.
22. **Hochstein, R. R., Thauberger, J. C., and Nyborg, E. O.,** Report on John Deere 3800 Forage Harvester, Bull. E0378B, Prairie Agricultural Machinery Institute, Portage la Prairie, Manitoba, 1979.
23. **Prince, R. P., Wheeler, W. C., and Fisher, D. A.,** Discussion on energy requirements for cutting forage, *Agric. Eng.,* 39(10), 638, 1958.

24. **Kepner, R. A., Bainer, Roy, and Barger, E. L.**, *Principles of Farm Machinery*, 2nd ed., AVI Publishing, Westport, Conn., 1972.
25. **Elfes, L. E.**, Design and development of a high-speed mower, *Agric. Eng.*, 35(3), 147, 1954.
26. **Colzani, G.**, Risultati di ricerche sperimentali eseguite su tre tipi di falciatrici operanti su foraggi diversi (Tests of Three Mowers on Different Forages), Rep. No. 33, Institute Sperimentale Di Meccanica Agraia, Milano, Italy, April, 1968, 30.
27. **Durfee, J. R., Johnson, C. E., and Turnquist, P. K.**, Eliptical cutter for forage harvesting, *Trans. ASAE*, 20(4), 635, 1977.
28. **Chancellor, W. J.**, An experiment on force and energy requirements for cutting padi stalks, *Malaysian Agric. J.*, 45(2), 200, 1965.
29. **Prince, R. P. and Wheeler, W. C.**, Factors Affecting the Cutting Process of Forage Crops, ASAE Paper No. 60-611, American Society of Agricultural Engineers, St. Joseph, Mich., 1960, 10.
30. **Rajput, D. S. and Bhole, N. G.**, Static and dynamic shear properties of paddy stem, *The Harvester*, Vol. 15, Agricultural Engineering Dept., Indian Institute of Technology, Kharagpur, W. B., India, 1973, 17.
31. **McRandal, D. M. and McNulty, P. B.**, Mechanical and physical properties of grasses, *Trans. ASAE*, 23(4), 816, 1980.
32. **Kepner, R. A.**, Analysis of the cutting action of a mower, *Agric. Eng.*, 33(11), 693, 1952.
33. **Dobler, K.**, Grundlegende Untersuchungen uber den freien Schnitt bei Halmgut, *Grundlagen Landtechnik*, 23(2), 1963.
34. **Dobler, K.**, Mehrkomponenten-Schnittkraftmessung mit Quarzkristollaufnehmern beim Mahen von Halmgut, *ATM Messtechnische Praxis*, Lieferung 447, R61, 1963.
35. **Dobler, K.**, Der freie Schnitt beim Mahen von Halmgut, *Hohenheimer Arbeiten*, Vol. 62, Eugen Ulmer, Stuttgart, 1972, 85.
36. **Feller, R.**, Effects of knife angles and velocities on cutting stalks without a counter-edge, *J. Agric. Eng. Res.*, 4(4), 277, 1959.
37. **McRandal, D. M. and McNulty, P. B.**, Impact cutting behavior of forage crops. I. Mathematical models and laboratory tests, *J. Agric. Eng. Res.*, 23(3), 313, 1978.
38. **McRandal, D. M. and McNulty, P. B.**, Impact cutting behavior of forage crops. II. Field tests, *J. Agric. Eng. Res.*, 23(3), 329, 1978.
39. **Lambert, M. B.**, Evaluation of Power Requirements and Blade Design for Slash Cutting Machinery, ASAE Paper No. 74-1570, American Society of Agricultural Engineers, St. Joseph, Mich., 1974.
40. **Bockhop, C. W. and Barnes, K. K.**, Power distribution and requirements of a flail-type forage harvester, *Agric. Eng.*, 36(7), 454, 1965.
41. **Light, R. G. and Yoerger, R. R.**, Power requirements for a horizontal rotary-type forage harvester, *Trans. ASAE*, 3(2), 89, 1960.
42. **Yukueda, M. and Kawamura, N.**, Studies of a Direct Throw Flail-Type Forage Harvester. I. On the Characteristics of Power Requirements and Chopped Lengths, Res. Rep. on Agric. Machinery, No. 7, Laboratory of Agricultural Machinery, Kyoto University, Japan, 1976, 32.
43. **van Rest, D. J.**, A harvest aid for sugar cane, *Agric. Eng.*, 51(3), 134, 1970.
44. **Alicbusan, L. C.**, The Rate of Substitution of Man Hours by Animal or Machine Horsepower Hours in Rice Production, Special Rep. Agricultural Engineering Dept., International Rice Research Institute, Los Banos Laguna, Philippines, August, 1964, 17.
45. **Teel, D. E.**, Developing a belt cutting system, *Proc. 1st Int. Grain Forage Harvesting Conf.*, American Society of Agricultural Engineers, St. Joseph, Mich., 1978, 233.
46. **Walker, J. T. and Bledsoe, B. L.**, A Band-Blade Header for Harvesting Soybeans, ASAE Paper No. 79-1585, American Society of Agricultural Engineers, St. Joseph, Mich., 1979.
47. **Chancellor, W. J.**, Design and construction of a chain mower, *Wis. Eng.*, 58, 14, 1954.
48. **Chancellor, W. J.**, Designing a chain mower, *ASAE Natl. Student J.*, p. 16, 1954.
49. **Bledsoe, B. L. and Porterfield, J. G.**, A balanced high-speed rotary sickle for cutting and trajecting plants, *Trans. ASAE*, 14(5), 818, 1971.
50. **Coates, W. E. and Porterfield, J. G.**, A compound helical cutterbar — design and field testing, *Trans. ASAE*, 18(1), 17, 1975.
51. **Miller, M. R.**, Developing a high capacity stalk cutter, *Agric. Eng.*, 49(3), 132, 1968.
52. **Hummel, J. W. and Winn, P. N., Jr.**, Measuring cutting forces of Maryland tobacco in relation to mechanical harvester design, *Trans. ASAE*, 10(1), 12, 1967.
53. **Bolen, J. S. and McColly, H. F.**, A rotary cutting mechanism for the direct harvest of dry edible beans, *Trans. ASAE*, 12(6), 862, 1969.
54. **Schertz, C. E.**, Rotary Cutters and Cylindrical Rotors for Gathering Function in Harvest of Soybeans, ASAE Paper No. 70-628, American Society of Agricultural Engineers, St. Joseph, Mich., 1970, 10.
55. **Hummel, J. W. and Nave, W. R.**, Impact Cutting of Soybean Plants, ASAE Paper No. 77-1555, American Society of Agricultural Engineers, St. Joseph, Mich., 1977.

56. **Moore, M. A., King, F. S., Davis, P. F., and Mamby, T. C. D.,** The effect of knife geometry on cutting force and fracture in sugar beet topping, *J. Agric. Eng. Res.,* 24(1), 11, 1979.
57. Urschel Co., How to Cut Food Products, Urschel Laboratories Inc., Valparaiso, Ind., 1977, 62.
58. **Schield, M. and Harriott, B. L.,** Cutting lettuce stems with a water jet, *Trans. ASAE,* 16(3), 440, 1973.
59. **Valco, T. D., Coble, C. G., and Ruff, J. H.,** High Pressure Water Jet Cutting of Sugar Cane, ASAE Paper No. 79-1569, American Society of Agricultural Engineers, St. Joseph, Mich., 1979.
60. **Akritidis, C. B.,** The mechanical characteristics of maize stalks in relation to the characteristics of cutting blade, *J. Agric. Eng. Res.,* 19(1), 1, 1974.
61. **Chancellor, W. J.,** Energy requirements for cutting forage, *Agric. Eng.,* 39(10), 633, 1958.
62. **Fischer-Schlemm, W. E.,** Der einfluss des Watenwinkles auf die Schneidhaltigkeit von Mahmesserklingen, *Grundlagen Landtechnik,* 5, 117, 1953.
63. **Fischer-Schlemmn, W. E. and Eggert, O.,** Der einfluss des Hacksel-messer Watenwinkles auf Schnitthaltigkeit und Kraftbedarf, *Landtechnische Forsch.,* 5(4), 109, 1955.
64. **Hennen, J. J. and Markham, D.,** Maintaining Sharp Forage Chopper Knives, ASAE Paper No. 71-681, American Society of Agricultural Engineers, St. Joseph, Mich., 1971, 5.
65. **McClelland, J. H. and Spielrein, R. E.,** A study of some design factors affecting the performance of mower knives, *J. Agric. Eng. Res.,* 3(2), 137, 1958.
66. **Prince, R. P. and Wheeler, W. C.,** Some Factors Affecting the Cutting of Forage Crops, Prog. Rep. No. 37, Agricultural Experiment Station, University of Connecticut, Storrs, 1959, 10.
67. **Richey, C. B.,** Discussion on energy requirements for cutting forage, *Agric. Eng.,* 39(10), 636, 1958.
68. **Schulze, K. H.,** Uber den Schneidvorgang an Grashalmen, *Grundlagen Landtechnik,* 5, 98, 1953.
69. **Stroppel, Th.,** Studien uber den Verschleiss von Schneiden fur Halmartiges Schnittgut, *Grundlagen Landtechnik,* 5, 134, 1953.
70. **Ige, M. T. and Finner, M. F.,** Forage harvester knife response to cutting force, *Trans. ASAE,* 19(3), 451, 1976.

AGRICULTURAL MACHINERY — FUNCTIONAL ELEMENTS — THRESHING, SEPARATING, AND CLEANING

Leroy K. Pickett and Neil L. West

INTRODUCTION

Separation is important for a wide variety of agricultural crops. Wheat may be separated from straw and chaff or potatoes may be separated from stones. Our purpose is to describe the machine elements for functions of threshing, separating, and cleaning as typically used when harvesting grain and seed crops. Threshing is detaching the seed from the head, cob, or pod; separating is isolating detached seed and small debris from the bulk of the straw; and cleaning is isolating the desired seed from chaff, small debris, and remaining unthreshed material.[1] Separation has the broader meaning of isolating something found in a combination that includes the functions of threshing, separating, and cleaning defined here. Separation effectiveness depends on the physical, mechanical, and aerodynamic properties of the grain and other materials. Crop properties are influenced by varietal characteristics, maturity, and moisture content.

Mechanisms used in grain harvesting machines may perform the functions of threshing, separating, and cleaning individually or in combination. In a typical machine arrangement, Figure 1, these functions are performed.

1. Crop material is fed into the threshing cylinder by a conveyor.
2. Grain is threshed by action of the rotating cylinder and stationary concaves.
3. Threshed material is conveyed by the threshing cylinder and beater to the straw walkers.
4. Grain is separated by the cylinder and beater grates and the straw walkers.
5. Grain is cleaned pneumatically and mechanically by fan air and action of oscillating sieves.
6. Grain from below the sieves is elevated to a temporary storage.
7. Material other than grain is discharged from the rear of the walkers and upper sieve.
8. Tailings material from the rear of the lower sieve is recycled to the threshing cylinder.

Material flow through harvesting machine processes of threshing, separating, and cleaning is presented in Figure 2. The illustration shows a crop with a relatively high material-other-than-grain (m.o.g.)-to-grain ratio, which is typical for cereal grains. Quantity of material by weight is indicated by the width of the flow streams.

THRESHING

The threshing process has the objective of severing the attachment of the grain kernels from the plant. In some crops, the kernel must also be removed from a protective husk or pod.

In a harvesting machine, the threshing process may be accomplished by impact between the grain and a fast moving element, by rubbing, by squeezing, or by a combination of these methods.

Threshing Mechanisms

The three types of threshing devices used in most of the present day combines are the transverse-flow cylinder with rasp bars, the transverse-flow cylinder with spike teeth, and the axial-flow cylinder with rasp bars. Impact between the threshing bars or teeth and grain

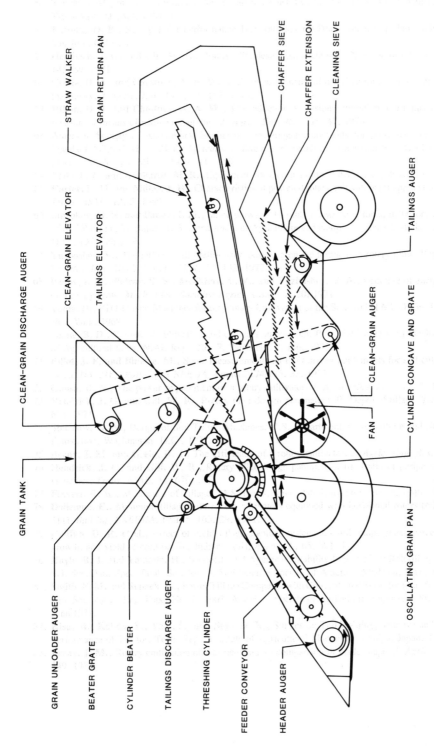

FIGURE 1.   Diagram of a grain harvesting machine.

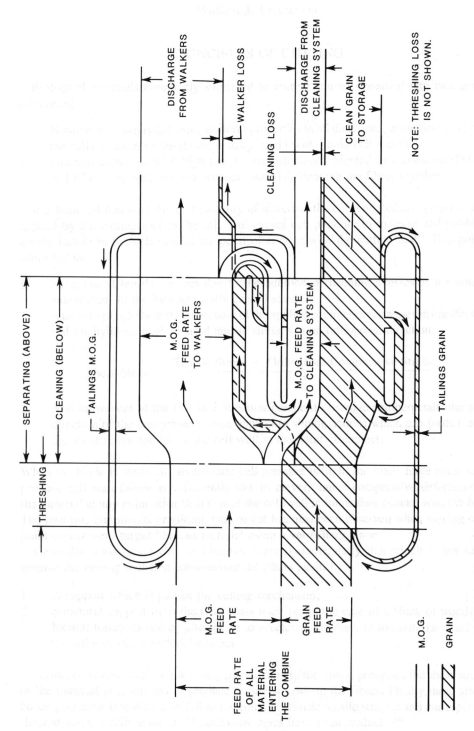

FIGURE 2.   Material flow through harvesting machine processes of threshing, separating, and cleaning.

create most of the threshing action, although rubbing undoubtedly contributes to the threshing process. These devices have all been used for a large number of crops with widely varying crop conditions.

Some devices have been built to provide rubbing and compression for threshing. A double-belt threshing unit was developed at the Oregon Agricultural Experimental Station in the late 1960s for harvesting grass seed and legumes. These crops have small, easily damaged seeds. Material was fed between two belts traveling in the same direction at different speeds. The clearance between the belts and the compressive force was controlled by two spring-loaded, back-up plates. A similar mechanism has also been used in laboratory tests to effectively thresh soybeans with minimal damage.[2] Rubber-covered steel rollers have also been used to thresh some podded crops.

Axial-flow caged shellers are used for threshing corn. A rotating cylinder with an auger, paddles, or lugs forces the ears of corn against each other as they move along the axis of the sheller. The peripheral cage, made of perforated metal or parallel bars, has openings for easy passage of shelled kernels but not the cobs. A large percentage of the threshing is accomplished by the relative motion and squeezing action of one ear of corn with another.

A laboratory roller sheller, using a rubber terra-tire, was developed for the purpose of reducing kernel damage during shelling of corn.[3] Kernel damage was substantially reduced in comparison to that for a transverse-flow cylinder with rasp bars. However, threshing efficiency for the roller sheller was lower than for the rasp-bar cylinder.

Threshing Principles

The force applied to a single kernel to sever its attachment point varies greatly, depending on the orientation of the applied force. If the attachment point is placed in pure tension, the applied force to sever a single kernel will be considerably greater than if a bending moment is applied. In some cases, a portion of the kernel may actually be used as a fulcrum for severing the attachment. Care must be taken, however, when analyzing the forces applied to an ear of grain that the kernels have space to move in the direction of the applied forces without causing damage to it or its neighboring kernels. The component of the applied force used to thresh the grain may be only a small portion of the total force.

With the impact and rubbing action of transverse- or axial-flow cylinders, the straw or stalks are subjected to many deformations, which absorb the majority of the energy. The energy efficiency of the threshing process is very low, especially with crops that have a high m.o.g.-to-grain weight ratio. The total energy required for the threshing process greatly depends on the condition and orientation of the crop material. In general, the energy is lower if the crop material is dry and oriented so that the heads are fed first into the threshing mechanism.

The rasp bar is, at present, the predominant threshing element used on both the transverse-flow and axial-flow cylinders or rotors. The number and profiles vary; but in general, they are a bar with a notched surface. Rasp bars are typically used in conjunction with a concave with a rough surface formed by a series of bars set at intervals parallel to the cylinder or rotor axis. Most concaves have open areas between the bars to allow quick separation of the threshed grain in order to minimize grain damage.

Spike-tooth cylinders, which were used almost exclusively until about 1930, are now limited to primarily rice and windrowed beans. The arrangement of a spike-tooth cylinder and concave is such that the teeth of the cylinder pass midway between staggered teeth on the concave, thus producing a combing action in addition to the high-speed impacts on the heads. The combing action of the spike-tooth cylinder generally produces more straw breakage than the compressive and rubbing action of a rasp-bar cylinder. The spike-tooth cylinder is less sensitive to uneven feeding and tough or wet straw conditions frequently experienced when harvesting rice. Power requirements when harvesting wheat with a spike-tooth cylinder were 75 to 85% of those with a rasp-bar cylinder.[4]

DESIGN FACTORS

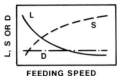 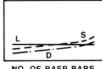 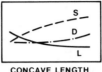

FEEDING SPEED   NO. OF RASP BARS   CONCAVE LENGTH   CYLINDER DIAMETER

ADJUSTMENTS AND OPERATION

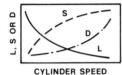 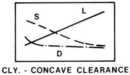 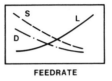

CYLINDER SPEED   CLY. - CONCAVE CLEARANCE   FEEDRATE

CROP CONDITIONS

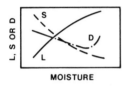 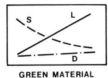

MOISTURE   GREEN MATERIAL

FIGURE 3.   General functional characteristics of a rasp-bar cylinder and open concave. L = unthreshed grain loss, D = grain damage, S = percent grain separated through concave. (From Wieneke, F., *Grundlagen Landtechnik*, 21, 33, 1964. With permission.)

A spike-tooth, concave grate typically has much less open area than a rasp-bar grate. In lab tests with rice, Neal and Cooper[5] found that a rasp-bar-cylinder concave separated 72% of the grain compared to only 52% by a spike-tooth grate with similar material and feed rates.

Axial-flow, threshing cylinders are similar to rasp-bar, transverse-flow cylinders except that the material flows in a spiral path around the cylinder instead of a relatively short path across an arc of the cylinder surface. Since the material makes multiple passes through the threshing zone, typically the tip speeds are slower, and the cylinder-to-concave clearances are wider than for a transverse system. Both of these conditions tend to reduce grain damage.

Threshing Performance Factors

Performance criteria important for threshing mechanisms include percent of grain threshed, percent of grain separated, and grain damage level. Wieneke[6] developed a generalized, graphical representation of selected performance characteristics for a rasp-bar cylinder with an open-grate concave based on the work of Arnold et al.,[7-9] and similar research in Germany by Wieneke and Caspers,[10] Baader,[11] and Schulze.[12] Figure 3 shows the qualitative relationships he found for nine factors influencing the threshing process.

These factors affecting threshing performance may be divided into two categories — those related to the design and adjustment of the machine components, and those related to the crop. Energy efficiency for the threshing process and peak power requirements for operating the mechanism are secondary factors that should be considered when evaluating threshing performance.

In crops such as the cereal grains, which have a high m.o.g.-to-grain ratio compared to crops with a low m.o.g.-to-grain ratio such as corn, m.o.g. feed rate is the dominant factor. Crop moisture content and amount of green material in the m.o.g. are examples of crop factors that have a large influence on threshing performance.

Another important crop related factor, but one that may be controlled by the design of the harvesting machine, is that of material orientation as it enters the threshing system. Heads first with heads on top of the straw mat is most common and one of the best orientations. Heads first but with heads on bottom has slightly better separation, but is very difficult to achieve with the typical harvesting machine configuration shown in Figure 1. The power requirements associated with a material orientation parallel to the cylinder axis becomes very feed-rate sensitive.[13]

Arnold and Lake[13] found that instantaneous power requirements depend more on the characteristics of the material feed stream and diameter of the threshing cylinder than on the average feed rate. Unevenness in the feed stream caused the power requirements to vary as much as an order of magnitude. Large-diameter, high-inertia cylinders have the advantage of storing sufficient energy for handling high peak power requirements. It was generally found that as cylinder diameter was increased from 300 mm (12 in.) to 525 mm (21 in.), the power requirements decreased. No further advantage was found for cylinder diameters greater than 525 mm.

Arnold and Lake[13] also found that for all but very low feed rates, less power is required to thresh a fast moving, thin stream of crop than a slow moving, thick one.

The following general statements can also be made concerning the mean power requirements of a threshing system:

1. As the cylinder speed increases, the power requirement increases linearly or slightly exponentially.
2. Increasing concave length increases power requirements.
3. Concave clearance has very little effect on power in the typical working range; however, power often increases with very small clearances.
4. A closed concave will require approximately 25% more power than an open concave.
5. A reduction in straw length will reduce power requirements and increase the separation rate.

SEPARATING

Separation begins as the threshed grain and chaff first leave the threshing area, which is through the concave for harvesters equipped with open grates. Many different mechanisms, including raddles, oscillating racks, straw walkers, and rotary separators, have been used to remove the grain remaining in the straw and coarse material discharged from the threshing cylinder. Walkers have been widely accepted for grain harvesting equipment built during the past 50 years. After many years of development, several machines introduced since 1975 are equipped with rotary separators.

Straw Walkers

A frequently used mechanism for separating grain and chaff from straw consists of a set of three to six straw walkers, each 8 to 12 in. wide. The walkers are mounted side-by-side on front and rear multiple-throw crankshafts. The crankshafts usually have 2- or 3-in. throws that are timed to give equal angular spacing for mounting the number of walkers used. The action of straw walkers accelerates the material up and toward the rear during a portion of the cycle and then lets it fall to the surface of the walker again. This loosens and agitates the material, allowing the grain to sift through and out the openings in the walker. It is difficult to completely describe the motion of material over straw walkers.[14]

Extensive laboratory studies on straw walker performance were conducted at the University of Saskatchewan.[15,16] Crop variables including straw length, straw moisture content, and crop variety and machine variables including straw walker length, speed, and crank-throw length were evaluated for a range of feed rates of the crop material. Straw length and moisture content had little effect on walker efficiency for wheat. However, walker losses were higher for dry oats and barley with moisture content approximately 7% (w.b.), than for tough straw conditions. Decreased walker efficiency for the dry straw condition was attributed to broken straw forming a dense mat, which was conveyed out by the walker with little agitation within the mat. Wheat straw did not form a mat even though the straw was severely broken.[16] Optimum efficiency for 2-in. throw walkers was at 200 rpm and for 3-in. throw walkers was at 150 rpm. When operated at optimum speed, efficiency for the 3-in. throw walkers was about the same as for the 2-in. throw walkers.

Rotary Separators

Interest in rotary threshing and separating of grain began with early efforts in mechanization of grain harvesting.[17] The concept of centrifugal threshing, separating, and cleaning prompted much discussion and research during the 1950s and 1960s.[18] Research in the late 1960s and into the 1970s was focused primarily on separation because it was not feasible to thresh with centrifugal force.[19,20] Long et al.[19] found that centrifugal separation was not effective unless the straw mass was agitated. Time for the seed to move through the straw mat was reduced as acceleration was increased to 35 times that of gravity. Using greater acceleration was not expected to significantly decrease separation time.

From 1975 to 1980, four manufacturers introduced combines in the U.S. with axial-flow rotary threshing and separation.[21,22] The rotor spirals the crop material toward the rear so that it passes the threshing concaves and separator grates several times. The material moves about one third the peripheral speed of the rotor rasp bars.[23] Centrifugal force assists separation through the open-grate threshing concaves and separator grates. Vanes inside the cage control the spiral movement of the material for effective separation. Rotor size varies from 17 in. in diameter by 88 in. long to 31.5 in. in diameter by 168 in. long.[22] Rotational speeds are from 198 rpm to 1565 rpm with smaller rotors having the higher speed ranges, and the larger rotors, the lower speed ranges.

CLEANING

The purpose of a combine cleaning system is to remove chaff and other small m.o.g. from the grain. The cleaning system usually has one or more oscillating sieves and a fan delivering air at the front and up through the sieves.

Sieving

Sieving is a process in which the material mixture is moved over a perforated surface with openings of specified shape and size. In a simple sieving system typically used for harvesting machines, the mixture is moved horizontally across a sieve with gravity accelerating particles through the openings. The material mixture is usually moved by oscillating the perforated surface, but may be assisted by pneumatic or other mechanical forces.

Efficiency of sieving is enhanced by oscillating the sieve at optimal frequency, direction, and amplitude. Movement of material on an oscillating surface was described by Turnquist and Porterfield[24] and Schertz and Hazen.[25] Sieve motion should satisfy the following three requirements in order to facilitate passage of particles through the openings: (1) bring particles into alignment with the openings, (2) provide conditions suitable for passage when particles are aligned, and (3) move particles to other openings if they do not pass through those tried.[26]

Particles that are caught in the openings reduce the efficiency of the sieve. Feller[27] found that clogging rates appear to be considerably higher for irregular shape, undersize particles than for oversize particles. Greater efficiency in sieving may be obtained by utilizing acceleration higher than that due to gravity. Park[28] developed a vertical rotating-sieve separator with an operational speed giving seven times the acceleration of gravity. Advantages cited for the separator were high capacity, insensitivity to slope, compactness, and ability to separate seeds from trashy materials. With multiple layers of material mixtures, interaction of particles is important for sieving. Vibration causes grain particles to act as a fluid resulting in segregation of the mixture. Tendency for this segregation may be used for particle positioning for improving separating efficiency. Particles with a combination of higher density and smaller diameter tend to sink, while particles with low density and larger diameter tend to rise.

Pneumatic Cleaning

Aerodynamic characteristics of particle mixtures are important for cleaning. Winnowing of grain was an ancient method of cleaning chaff from grain. Though seldom effective alone in harvesting machines, assistance of air is used for most cleaning mechanisms. Air is utilized to remove light materials from the mixture, to assist in positioning particles over sieve openings, and to move particles along the surface if they do not pass through the openings. A series of studies on aerodynamic behavior using oscillating sieves was conducted at the University of Guelph in Ontario, Canada.[29-34] Performance of separating grain from straw and chaff was evaluated for an oscillating shoe similar to those typically used for cleaning in combine harvesters. At low feed rates of wheat, aerodynamic separation of grain from straw and chaff took place over the sieve. Optimum settings determined were 36° sieve lip angle and 330 to 380 c/min oscillation frequency. When feed rate was increased, the material particles were no longer supported aerodynamically, which caused a mat to form on the sieve. As the mat formed, grain loss over the sieve increased.[29] Dropping the material mixture through an air stream before it entered the front of the sieve dispersed the mat. Further increases in material feed rate resulted in aerodynamic suspension of material at the front of the sieve and the forming of a mat toward the rear.[32] Propelling the mixture downward at a higher rate of speed than delivered by free fall allowed higher air stream velocities without excessive aerodynamic grain loss and permitted much greater feed rates while maintaining aerodynamic separation.[33] For effective separation, it was necessary to increase sieve lip angle to 45° and decrease the oscillation frequency to 250 c/min. Because of differences in the physical properties of the grain, chaff and straw, aerodynamic separation was not as effective for barley as for wheat.[34]

With aerodynamic separation at the front of the sieve, chaff and straw are moved to the rear. MacAulay and Lee[32] reported that two thirds of the lower sieve served no useful purpose because the foreign material passing through the upper sieve was carried by the air stream to the rear third of the lower sieve. This principle was utilized for a cascade system of three sieves where the sieve immediately below is located about half its length rearward of the sieve above.[35]

Avoidance of grain loss with the straw and chaff discharge is usually of primary concern in separation effectiveness. A second concern is elimination of debris from the grain. Frequently, settings which reduce debris in the grain produce the undesired effect of increased grain losses.[36-40]

PERFORMANCE OF SYSTEMS FOR THRESHING, SEPARATING, AND CLEANING

Performance evaluation has been critical in the development and use of grain harvesters.

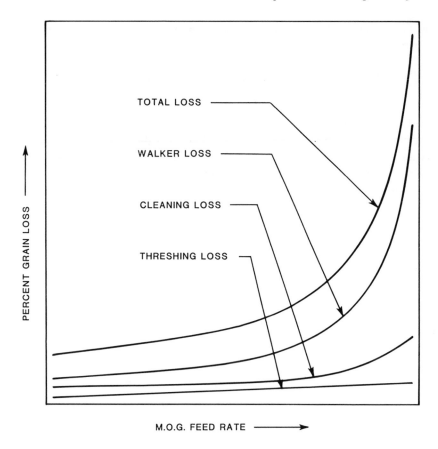

FIGURE 4. Typical relationships for grain losses from combine efficiency studies.

Although no substitute has been found for field performance tests, laboratory tests have advanced development substantially. Efficiency of saving seed and avoidance of mechanical damage during threshing are the main purposes for conducting performance tests. Many performance tests have been conducted by manufacturers during the development of new combines, particularly when a new principle is incorporated in the design.[37] Researchers at educational institutions have reported on field performance evaluations for a number of commercially important grain and seed crops in the U.S. including wheat, corn, barley, grain sorghum, and edible beans.[38-42] For efficiency studies, grain losses are usually plotted vs. m.o.g. feed rate. Figure 4 shows typical relationships for harvesting small grains. Conditions in which straw breaks up, such as dry barley, will cause the slope of the separating loss curve to increase more rapidly. Reducing threshing aggressiveness will usually decrease the slope of the separating loss curve but increase the slope of the threshing loss curve. Very finely broken straw will likely increase cleaning system load and cause the slope of the cleaning loss curve to increase.

Nyborg et al.[43] analyzed data collected for a series of tests over a range of feed rates for wheat, barley, oats, and rye. They found that walker losses were described by the equation:

$$L = a(M)^b \times (R)^c$$

where L = percent loss, M = m.o.g. feed rate, R = Grain-to-m.o.g. ratio, and a,b,c, are constants for crop conditions.

A simple linear relationship between percent grain loss and feed rate gave a better fit for cleaning and threshing losses except for very high feed rates, for which the equation given above gave a better fit.

PHYSICAL PROPERTIES OF CROPS ASSOCIATED WITH THRESHING, SEPARATING, AND CLEANING

In order to better understand and design grain harvesting equipment, physical properties of the crops being processed are often very helpful. For this reason the following table of properties (Table 1) compiled from the published literature cited is included. Since a successful harvesting machine must harvest essentially 100% of the crop available and not just an "average" seed, a range of values for each property was given when available.

Table 1
PHYSICAL CROP PROPERTIES

Property	Wheat	Barley	Oats	Rye	Corn	Soybeans	Sorghum	Rice	Ref.
Aerodynamic properties									
Drag Coefficient **a**									
Broken corn cobs					1.40—4.50				44
Grain	0.50—2.40	0.500—0.570	0.470—0.640		0.35—3.00	0.45—1.50		3.0—4.0	44—49
Straw									44, 45, 50, 51
With node at one end									
(0.5—7.0 in.)		0.616—0.910	0.710—1.000						
With node in the middle									
(0.5—7.0 in.)	0.66—0.94	0.553—1.048	0.927—1.191						
(0.25—10.0 in.)			0.720—0.870						
Without node									
(0.5 in.)	2.00—5.00								
(1.0 in.)	2.80—6.50								
(0.5—7.0 in.)		0.750—1.230	0.839—1.143	0.68—0.86	0.66—1.12	0.30—0.66			
Resistance coefficient ($lbs\text{-}sec^2/ft^2 \times 10^{-6}$) **b**									
Chaff	0.412								
Grain	0.112—0.124	0.104—0.162	0.088—0.163		0.626	0.298			51
Straw									
With node at one end									
(0.5—7.0 in.)	0.197—0.786	0.395—0.553	0.593—0.830						45, 51, 52
(0.25—10.0 in.)									
Without node and node in the middle									
(0.5—7.0 in.)	0.251—1.003	0.575—0.805	0.835—1.169						45, 51
(0.25—10.0 in.)									
Threshed heads **c**	8.954								51
Reynolds number **c**									
At terminal velocity									
Grain	2720	2280—2610	1758—2480		5770	6280			45, 46
Terminal velocity (ft/sec) **d**									
Chaff	3.3—8.3								51, 53
Grain	18.3—41.0	16.7—33.0	15.0—29.0	20.0—27.0	26.0—51.0	30.0—60.0	31.0—45.0	17.0—23.0	45, 46, 49—54
Silage (Up to 5.0 in.)					5.9—18.8				55
Stalk (2.7—4.0 in.)						12.0—22.0			50
Stover **e** (0.3—4.2 in.)					20.0—55.0				50

Table 1
PHYSICAL CROP PROPERTIES (continued)

Property	Wheat	Barley	Oats	Rye	Corn	Soybeans	Sorghum	Rice	Ref.
Straw									
Mixture of straw **f** (0.5—7.0 in.)	6.0—18.0	6.7—23.3	6.7—20.0	8.0—18.0					45, 50, 51, 53
With node **g** (Up to 0.5 in.)	13.3—15.0								
With node at one end (0.25—10.0 in.)	9.0—16.0	6.7—10.15	8.6—11.0						
With node in the middle (0.25—10.0 in.)	8.3—15.8	6.6—11.6	6.9—11.3						
Without node (Up to 0.5 in.)	8.3—10.0								
(0.25—10.0 in.)	8.1—10.5	5.70—7.25	7.2—8.4						
Threshed heads	7.09								51
Mechanical properties									
Angle of flow (degree) **h** — Grain	36.5—45.0	37.0—42.5			34.0—44.0		37.5—39.0	34.0—36.0	56, 57
Angle of rebound (degree) **i** — Grain (On steel plate)	34.5—58.2				36.7—70.9	21.2—64.1			58
Angle of repose (degree) — Angle of repose — Kinetic **j** — Grain / Angle of repose	16.0	16.0	18.0	17.0	16.0	16.0	20.0	20.0	59
Static **k** grain	27.0—41.0	28.0—34.0	30.0—32.0	26.0—30.0	27.0—38.0	29.0—30.0	30.0—45.0	30.0—45.0	57, 59, 60, 61, 62, 63
Coefficient of friction **l** — Coefficient of friction — Kinetic **m** grain (On aluminum)	0.350—0.610								
(On grain)	0.390—0.630				0.226—0.280			0.950—1.025	43, 63—68
(On plywood)	0.310—0.570								
(On stainless steel)									
(On steel)	0.200—0.525								
Coefficient of friction — Static **n** grain		0.376—0.425	0.404	0.40	0.259—0.309	0.404	0.70	0.556—0.606	56, 57, 59, 61—63, 69—71

									Ref.
(On concrete)	0.36—0.69	0.23—0.62	0.280—0.650	0.350—0.850	0.270—0.680	0.25—0.55	0.33	0.46—0.61	
(On grain)	0.47—0.53	0.51—0.53	0.530—0.620	0.490	0.510—0.520	0.55	0.65	0.68—0.73	
(On metal)	0.10—0.55	0.17—0.40	0.180—0.440	0.406—0.410	0.200—0.760	0.18—0.45	0.29—0.52	0.40—0.45	
(On plastic)	0.12—0.45	0.11—0.35	0.110—0.500		0.120—0.380	0.17—0.43			
(On wood)	0.25—0.59	0.21—0.50	0.200—0.500	0.370—0.550	0.240—0.380	0.29—0.45	0.30—0.51	0.44—0.61	
Silage									71
(On concrete)			0.456—0.699						
(On metal)			0.493—0.569						
(On plastic)			0.184—0.401						
(On wood)			0.563—0.583						
Straw									71, 73
(On concrete)	0.22	0.202—0.454							
(On metal)		0.304—0.351							
(On plastic)		0.139—0.219							
(On wood)		0.197—0.260							
Detachment force (lbs/kernel) o									
Radial force p									
Grain				1.64—7.10					74—77
Grain (compressed) q				2.20—4.50					76
Tangential force r									
Grain				1.20—2.55					76
Grain (compressed) q				0.80—1.80					76
Strength (lbs)									
(per grain, straw, etc.)									
Breaking strength s									
Cob (per in. of length)					3.6—42.2				82
Grain	14.77—31.09			77.0—313.0					74, 75
Straw	0.388—1.497			24.9—225.8					74, 75, 78—80
Physical properties									
Area (in.²)									
Area in least cross section									
Grain	0.0062—0.0091	0.0034—0.0045	0.0051—0.0066		0.050—0.061	0.024—0.047			50
Cross sec. area of									
eq. vol. sphere									
Broken corn cobs			0.01699—0.04579						44
Grain	0.00913—0.01728		0.06311—0.09672			0.03269—0.07142			44
Straw with node									44
(.050 in.)	0.01728—0.02477								
(0.75 in.)	0.02045—0.03182								
(1.00 in.)	0.02606—0.03874								

Table 1
PHYSICAL CROP PROPERTIES (continued)

Property	Wheat	Barley	Oats	Rye	Corn	Soybeans	Sorghum	Rice	Ref.
Frontal area									
Grain		0.03456	0.0193—0.3125						45
$(\pi/4 \times L_1 \times L_2 \times \cos\theta)$ t									
Straw $(L \times D)$ u									45, 51
With node at one end									
(.50—7.0 in.)	0.0380—1.5300	0.06264—0.81346	0.08957—1.23754						
(.25—10.0 in.)									
With node in the middle									
(.50—7.0 in.)	0.0383—1.4832	0.06624—1.02614	0.0810—1.28304						
(.25—10.0 in.)									
Without node									
(.50—7.0 in.)	0.0415—1.5264	0.0720—0.89107	0.0947—1.20672						
(.25—10.0 in.)									
Maximum projected area									
Chaff								0.04—0.12	83
Grain	0.015—0.023	0.021—0.037	0.022—0.041	0.013—0.023	0.0816—0.1998	0.039—0.067			74, 75, 84
Surface area									
Grain								0.0623—0.1008	48, 85
Surface area of equal volume sphere									
Grain								0.561—0.0610	48
Density (lbs/ft³)									
Absolute density									
Grain	62.4—92.8	83.1—88.7	46.1—86.1	72.4—76.1	71.1—82.2	72.4—75.5	82.0—82.2	82.67—86.60	50, 54, 80, 85—90
Bulk density									
Chaff	2.91—3.09	2.03—2.15							91
Grain	35.60—53.60	35.00—50.07	22.133—36.000	39.3—48.0	33.76—51.20	43.6—50.0	32.0—52.0	32.0—48.0	v
Hulls			8.0					20.0	60
Straw	0.854—2.130	1.29—1.35							91, 92
Straw and chaff mixture	2.570—2.690	2.81—2.97							91
Diameter (in.)									
Arithmetic mean (a + b + c)/3									
Grain	0.159—0.186	0.205—0.234	0.211—0.268		0.647	0.276	0.147	0.177—0.189	80, 90
Cross section diameter									
Ear					1.7—2.3				75
Straw	0.05—0.166	0.121—0.150	0.162—0.210						44, 45, 51
Diameter of equal volume sphere									
$(6v/\pi)^{1/3}$									
Broken corn cobs					0.4661—0.7637				44

	1	2	3	4	5	6	7	8	Ref.
Grain	0.1078—0.1880	0.145—0.210	0.138—0.253		0.2500—0.5000	0.2039—0.3019	0.137	0.128—0.136	44, 45, 90, 93, 94, 44
Straw with node									
(0.50 in.)	0.1448—0.1776								
(0.75 in.)	0.1608—0.2016								
(1.00 in.)	0.1824—0.2220								
Geometric mean (abc)$^{1/3}$									80, 90
Grain	0.152—0.177	0.169—0.199	0.158—0.189		0.636	0.275	0.144	0.1460—0.1620	
Intermediate diameter (b)									46, 51, 54, 80, 84, 85, 90, 95
Grain	0.090—0.157	0.090—0.173	0.078—0.169	0.071—0.105	0.261—0.676	0.221—0.286	0.146—0.169	0.0945—0.1530	
Major diameter (a)									w
Grain	0.201—0.323	0.265—0.508	0.315—0.720	0.217—0.313	0.382—0.840	0.240—0.330	0.150—0.185	0.2800—0.4415	
Minor diameter (c)									x
Grain	0.082—0.156	0.073—0.130	0.052—0.130	0.070—0.101	0.136—0.532	0.174—0.251	0.0829—0.130	0.0748—0.1100	
Seed count (seeds/lb)									54, 90
Grain	8,340—14,600	7,840—13,250	10,980—16,300		1,260—1,340	14,800—16,800		15,060—22,300	
Volume (in.^{3} × 10^{-3})									
Particle volume									48, 85, 90, 95
Grain	1.3936—2.2398	1.5482—2.4189	1.3025—1.783		13.43—17.21	1.2905—1.3775		0.98—1.44	
Weight									
Particle weight (lb × 10^{-3})									
Broken corn cobs					0.4401—2.866				44
Chaff	0.007							0.02293—0.06878	51, 83
Grain	0.0287—0.11576	0.073—0.173	0.0293—0.1239	0.0392—0.0529	0.5027—1.0648	0.06135—0.06549		0.04850—0.06504	y
Straw					0.2161—0.646				44, 45, 51
Short straw with node									
(.50 in.)	0.0154—0.0529								
(.75 in.)	0.0198—0.0772								
(1.00 in.)	0.0287—0.1036								
With node at one end									
(.50—7.0 in.)	0.0677—0.8165	0.0418—0.2898	0.0871—0.5849						
(.25—10.0 in.)									
With node in the middle									
(.50—7.0 in.)	0.0652—0.7466	0.0518—0.4035	0.0908—0.6033						
(.25—10.0 in.)									
Without node									
(.50—7.0 in.)	0.0223—0.7393	0.02025—0.2217	0.0473—0.5359						
(.25—10.0 in.)									
Threshed heads	0.4501								51
Weight per bushel (lb)									
Grain	60	48	32	56	56	60	50, 56	45	61

Notes: **a** Drag Coefficient — A factor representing the ratio of the aerodynamic drag acting on an object to the product of the airspeed and the area of the object. (A dimensionless number.)

b Resistance Coefficient — Is the weight of the particle divided by the square of the terminal velocity.

c Reynolds Number — An abstract number characteristic of the flow of a fluid past an obstruction. A ratio of inertia forces to viscous forces.

Table 1
PHYSICAL CROP PROPERTIES (continued)

d   Terminal Velocity — The listing uniform velocity attained by a moving body when the resistance of the air has become equal to the force of gravity.

e   Stover — Cobs, grain and stalks mixed.

f   Straw mixture consisting of straw with nodes and straw without nodes.

g   Node location not specified.

h   Angle of Flow — Angle of the material's surface with horizontal at which the flow begins.

i   Angle of Rebound — Angle that the rebounding seed made with the vertical. (Depends on height of fall, moisture content of particles, and thickness of plate.)

j   Angle of Repose, Kinetic — The angle of repose (angle with horizontal) in filling or piling of material.

k   Angle of Repose, Static — The angle of repose (angle with horizontal) in emptying or funneling, also angle of the material at rest.

l   Coefficient of Friction — Ratio between the force of friction and the force normal to the surface of contact.

m   Coefficient of Friction, Kinetic — Ratio between the force of friction and the force normal to the surface of contact for a system where the sources are in relative motion.

n   Coefficient of Friction, Static — Ratio between the force of friction and the force normal to the surface of contact for a system where surfaces are at rest.

o   Detachment Force — Maximum tensile strength of the rachilla which connects the kernel to the cob.

p   Radial Force — Force required to pull kernels radially from the cob.

q   Compressed Grain — Ears of corn which had been put under 250 lb of force, before detachment test. Each row of kernels had been put under equal force.

r   Tangential Force — Force required to pull kernels in the direction tangent to the circumference of the ear.

s   Breaking Strength — Also called maximum strength, destructive force, or rupture point.

t L_1 and L_2 are the average length and width and Θ is the orientation angle relative to the horizontal plane of the kernel in the air stream.

u $L \times D$ = length times diameter.

v   References: 48, 50, 54, 56, 57, 60—62, 66, 68, 85—87, 90, 91, 94, 96—99, 100—103.

w   References: 46, 48, 52, 54, 80, 84, 85, 88, 90, 95, 96.

x   References: 46, 48, 51, 54, 80, 84, 85, 90, 95, 96.

y   References: 44, 46—48, 50, 52, 54, 87, 90, 95, 104, 105.

REFERENCES

1. Terminology for Combines and Grain Harvesting, ASAE Standard: ASAE S343, American Society of Agricultural Engineers, St. Joseph, Mich., 1977.
2. **Walker, D. S. and Schertz, C. E.,** Soybean threshing by the relative motion of parallel belts, ASAE Paper No. 70-631, American Society of Agricultural Engineers, St. Joseph, Mich., 1970.
3. **Brass, R. W. and Marley, S. J.,** Roller sheller: low damage corn shelling cylinder, *Trans. ASAE,* 16, 64, 1973.
4. **Dodds, M. E.,** Power requirements of a self-propelled combine, *Can. Agric. Eng.,* 10, 74, 1968.
5. **Neal, A. E. and Cooper, G. F.,** Performance testing of combines in the lab, *Agric. Eng.,* 49, 397, 1968.
6. **Wieneke, F.,** Performance characteristics of the rasp bar thresher, *Grundlagen Landtechnik,* 21, 33, 1964.
7. **Arnold, R. E.,** Experiments with rasp bar threshing drums. I. Some factors affecting performance, *J. Agric. Eng. Res.,* 9, 99, 1964.
8. **Arnold, R. E., Caldwell, F., and Davis, A. C. W.,** The effect of moisture content of the grain and the drum setting of the combine-harvester on the quality of oats, *J. Agric. Eng. Res.,* 3, 336, 1958.
9. **Arnold, R. E. and Lake, J. R.,** Experiments with rasp bar threshing drums. II. Comparison of open and closed concaves, *J. Agric. Eng. Res.,* 9, 250, 1964.
10. **Wieneke, F. and Caspers, L.,** Influence of feeding speed, cylinder tip speed, clearance, and green material on the threshing process with different crops, *Grundlagen Landtechnik,* 21, 7, 1964.
11. **Baader, W.,** Influence of the feeding direction, position of feeding point to the cylinder, and position of threshing bars on the cylinder on the threshing process, *Grundlagen Landtechnik,* 21, 16, 1964.
12. **Schulze, K. H.,** Research on the threshing process with different types of rasp bar cylinders, *Grundlagen Landtechnik,* 21, 30, 1964.
13. **Arnold, R. E. and Lake, J. R.,** Experiments with rasp bar threshing drums. III. Power requirement, *J. Agric. Eng. Res.,* 9, 348, 1964.
14. **Kanafojski, Cz. and Karwowski, T.,** Agricultural machines theory and construction, in *Crop-Harvesting Machines,* Vol. 2, 1972, chap. 8. (Translated by the Foreign Scientific Publications Department of the National Center for Scientific, Technical and Economic Information, Warsaw, 1976.)
15. **Reed, W. B., Zoerb, G. C., and Bigsby, F. W.,** A laboratory study of grain-straw separation, *Trans. ASAE,* 17, 452, 1974.
16. **Reed, W. B. and Zoerb, G. C.,** A Laboratory Study of Straw Walker Efficiency, ASAE Paper No. 72-638, American Society of Agricultural Engineers, St. Joseph, Mich., 1972.
17. **Quick, G. R. and Buchele, W. F.,** *The Grain Harvesters,* American Society of Agricultural Engineers, St. Joseph, Mich., 1978, chap. 23.
18. **Lamp, B. J. and Buchele, W. F.,** Centrifugal threshing of small grains, *Trans. ASAE,* 3, 24, 1960.
19. **Long, J. D., Hamdy, M. Y., and Johnson, W. H.,** Centrifugal force and wheat separation, *Agric. Eng.,* 50, 578, 1969.
20. **Srivastava, A. K., Hamdy, M. Y., Nelson, G. L., Roller, W. L., and Huber, S. G.,** Centrifugal grain-straw separation. I. Theoretical analysis. II. Experimental investigation, *Trans. ASAE,* 17, 198, 1974.
21. **Rowland-Hill, E. W.,** Twin Rotor Combine Harnesses Potential of Rotary Threshing and Separation, ASAE Paper No. 75-1580, American Society of Agricultural Engineers, St. Joseph, Mich., 1975.
22. Red Book, *Implement Tractor,* 95, B-1, 1980.
23. **Murray, D. A.,** Progress in Crop Harvesting Equipment, SAE Paper No. 770706, Off-Highway Vehicle Meeting of Society of Automotive Engineers, Milwaukee, 1977.
24. **Turnquist, P. K. and Porterfield, J. G.,** Four-bar links for separating and conveying, *Trans. ASAE,* 4, 188, 1961.
25. **Schertz, C. E. and Hazen, T. E.,** Predicting motion of granular material on an oscillating conveyor, *Trans. ASAE,* 6, 6, 1963.
26. **Feller, R. and Foux, A.,** Oscillating screen motion effect on the particle passage through perforations, *Trans. ASAE,* 18, 926, 1975.
27. **Feller, R.,** Clogging rate of screens as affected by particle size, *Trans. ASAE,* 20, 758, 1977.
28. **Park, J. K.,** Vertical Rotating Screens for Separating Seeds from Trashy Materials, ASAE Paper No. 72-639, American Society of Agricultural Engineers, St. Joseph, Mich., 1972.
29. **Lee, J. H. A. and Winfield, R. G.,** Influence of oscillating frequency on separation of wheat on a sieve in an airstream, *Trans. ASAE,* 12, 886, 1969.
30. **German, R. F. and Lee, J. H. A.,** Grain separation on an oscillating sieve as affected by air volume and frequency, *Trans. ASAE,* 12, 883, 1969.
31. **Bilanski, W. K. and Dongre, S. P.,** Transportation of Wheat Grain Along the Combine Shoe, ASAE Paper No. 67-150, American Society of Agricultural Engineers, St. Joseph, Mich., 1967.
32. **MacAulay, J. T. and Lee, J. H. A.,** Grain separation on oscillating combine sieves as affected by material entrance conditions, *Trans. ASAE,* 12, 648, 1969.

33. **Rumble, D. W. and Lee, J. H. A.,** Aerodynamic separation in a combine shoe, *Trans. ASAE,* 13, 6, 1970.

34. **Misener, G. C. and Lee, J. H. A.,** Aerodynamic separation of grain from straw and chaff in a dispersed stream, *Can. Agric. Eng.,* 15, 62, 1973.

35. **Gordon, L. L., Sumsion, F. D., and Robinson, T. S.,** Development of the MF Cascade Shoe, ASAE Paper No. 69-621, American Society of Agricultural Engineers, St. Joseph, Mich., 1969.

36. **Simpson, J. B.,** Effect of longitudinal slope on combine-shoe performance, *Trans. ASAE,* 9, 1, 1966.

37. **Mark, A. H., Godlewski, K. J. M., and Coleman, J. L.,** A Global Approach to the Testing and Evaluation of Combine Performance, ASAE Paper No. 62-126, American Society of Agricultural Engineers, St. Joseph, Mich., 1962.

38. **Goss, J. R., Kepner, R. A., and Jones, L. G.,** Performance characteristics of the grain combine in barley, *Agric. Eng.,* 39, 697, 1958.

39. **Johnson, W. H.,** Efficiency in combining wheat, *Agric. Eng.,* 40, 16, 1959.

40. **Waelti, H., Buchele, W. F., and Farrell, M.,** Progress report on losses associated with corn harvesting in Iowa, *J. Agric. Eng. Res.,* 14, 134, 1969.

41. **Pickett, L. K.,** Mechanical damage and processing loss during navy bean threshing, *Trans. ASAE,* 16, 1047, 1973.

42. **Fairbanks, G. E., Johnson, W. H., and Shrock, M. D.,** Grain Sorghum Harvesting Loss Study, ASAE Paper No. 76-1556, American Society of Agricultural Engineers, St. Joseph, Mich., 1976.

43. **Nyborg, E. O., McColly, H. F., and Hinkle, R. T.,** Grain-combine loss characteristics, *Trans. ASAE,* 12, 727, 1969.

44. **West, N. L.,** Aerodynamic Force Predictions, ASAE Paper No. 70-334, American Society of Agricultural Engineers, St. Joseph, Mich., 1970.

45. **Bilanski, W. K. and Lal, R.,** Aerodynamic Properties of Threshed Oats and Barley Straw and Grain, ASAE Paper No. 66-610, American Society of Agricultural Engineers, St. Joseph Mich., 1966.

46. **Bilanski, W. K., Collins, S. H., and Chu, P.,** Aerodynamic properties of seed grains, *Agric. Eng.,* 43, 216, 1962.

47. **Garrett, R. E. and Brooker, D. B.,** Aerodynamic drag of farm grains, *Trans. ASAE,* 8, 49, 1965.

48. **Chand, P. and Ghosh, D. P.,** Dynamics of particles under pneumatic conveyance, *J. Agric. Eng. Res.,* 13, 27, 1968.

49. **Hawk, A. L., Brooker, D. B., and Cassidy, J. J.,** Aerodynamic characteristics of selected farm grains, *Trans. ASAE,* 9, 48, 1966.

50. **Uhl, J. B. and Lamp, B. J.,** Pneumatic Separation of Grain and Straw Mixtures, ASAE Paper No. 61-135, American Society of Agricultural Engineers, St. Joseph, Mich., 1961.

51. **Bilanski, W. K. and Lal, R.,** Behavior of threshed materials in a vertical wind tunnel, *Trans. ASAE,* 8, 411, 1965.

52. **Bilanski, W. K., Collins, S. H., and Chiu, C. L.,** The behavior of seed grains in a vertical wind tunnel, *Can. Agric. Eng.,* 5, 1963.

53. **Persson, S.,** Properties of shoe materials in combines, *Landtechnische Forsch.,* Munich, Vol. 7, 1957.

54. **Goss, J. R.,** Some Physical Properties of Forage and Cereal Crop Seeds, ASAE Paper No. 65-813, American Society of Agricultural Engineers, St. Joseph, Mich., 1965.

55. **Wolfe, R. R. and Tatepo, C. G.,** Terminal Velocity of Chopped Forage Material, ASAE Paper No. 70-362, American Society of Agricultural Engineers, St. Joseph, Mich., 1970.

56. **Kramer, H. A.,** Factors influencing the design of bulk storage bins for rough rice, *Agric. Eng.,* 25, 466, 1944.

57. **Lorenzen, R. J.,** Effect of Moisture Content on Mechanical Properties of Small Grain, Master's thesis, University of California, Davis, 1957.

58. **Sharma, R. K.,** Rebounding Characteristics of Grains Falling on Steel Surfaces, Thesis, University of Guelph, Ontario, Canada, 1969.

59. **Stahl, B. M.,** Grain Bin Requirements, USDA Circular 835, U.S. Department of Agriculture, Washington, D.C., 1950.

60. **Mohsenin, N. N.,** Physical properties of plant and animal materials, *Texture of Foods, Mechanical Damage, Aero- and Hydrodynamic Characteristics and Frictional Properties,* Vol. 1, First Preliminary Edition (Part 2), March 1968.

61. *ASAE Yearbook,* American Society of Agricultural Engineers, St. Joseph, Mich., 1973.

62. **Ketchum, M. S.,** *The Design of Walls, Bins, and Grain Elevators,* McGraw-Hill, New York, 1919.

63. **Long, J. D.,** Design of grain storage structures, *Agric. Eng.,* 12, 274, 1931.

64. **Zoerb, G. C.,** Physical Properties of Wheat for Moisture Content Determination, ASAE Paper No. 70-363, American Society of Agricultural Engineering, St. Joseph, Mich., 1970.

65. **Snyder, L. H., Roller, W. L., and Hall, G. E.,** Coefficient of Kinetic Friction of Wheat on Various Metal Surfaces, ASAE Paper No. 65-321, American Society of Agricultural Engineers, St. Joseph, Mich., 1965.

66. **Bickert, W. G. and Buelow, F. H.,** Kinetic friction of grains on surfaces, *Trans. ASAE*, 9, 129, 1966.
67. **Bickert, W. G. and Buelow, F. H.,** Some coefficients of friction of grains sliding on surfaces, *Q. Bull. Mich. Agric. Exp. Stn.*, 47(3), 1965.
68. **Sharan, G.,** Stresses in Moving Grain, Master's thesis, University of Guelph, Ontario, Canada, 1967.
69. **Brubaker, J. E. and Pos, J.,** Determining static coefficient of friction of grains on structural surfaces, *Trans. ASAE*, 8, 53, 1965.
70. **Airy, W.,** The pressure of grains, *Inst. Civil Eng. Min. Proc.*, 121, 347, 1898.
71. **Jamieson, J. A.,** Grain pressures in deep bins, Canadian Society of Civil Engineers, *Transactions 1093*, 17(2), 554, 1905.
72. Farm Building Standards, Suppl. No. 6 to the National Building Code of Canada.
73. **Richter, D. W.,** Friction coefficients of some agricultural products, *Agric. Eng.*, 35, 411, 1954.
74. **Waelti, H. and Buchele, W. F.,** Factors affecting corn kernel damage in combine cylinders, *Trans. ASAE*, 12, 55, 1969.
75. **Waelti, H.,** Physical Properties and Morphological Characteristics of Maize and Their Influence on Threshing Injury of Kernels, Ph.D. thesis, Iowa State University, Ames, 1967.
76. **Fox, R. E.,** Development of a Compression Type Corn Threshing Cylinder, Master's thesis, Iowa State University, Ames, 1969.
77. **Burmistrova, M. F. et al.,** Physicomechanical Properties of Agricultural Crops, Israel Program for Scientific Translations, Jerusalem, 1963, 32.
78. **Zoerb, G. C. and Hall, C. W.,** Some mechanical and rheological properties of grains, *J. Agric. Eng. Res.*, 5, 83, 1960.
79. **Shpolyanskaya, A. L.,** Structural-mechanical properties of the wheat grain, *(NAIE Translation 169)*, *Colloid J.*, *(USSR)*, 14, 137, 1952.
80. **Bacchus, N. N.,** The Mechanical Properties of Soybeans, Master's thesis, University of Guelph, Ontario, Canada, 1971.
81. **Middendorf, F. J.,** Physical and Mechanical Properties of Rapeseed, Master's thesis, University of Guelph, Ontario, Canada, 1972.
82. **Salmon, S. C.,** An instrument for determining the breaking strength of straw, and a preliminary report on the relation between breaking strength and lodging, *J. Agric. Res.*, 43, 73, 1931.
83. **Kashayap, M. M. and Pandya, A. C.,** Air velocity requirements for winnowing operations, *J. Agric. Eng. Res.*, 11, 24, 1966.
84. **Edison, A. R. and Brogan, W. L.,** Size Measurement Statistics of Kernels of Six Grains, ASAE Paper No. 72-841, American Society of Agricultural Engineers, St. Joseph, Mich., 1972.
85. **Wratten, F. T., Poole, W. D., Chesness, J. L., Bal, S., and Ramarao, V.,** Physical and Thermal Properties of Rough Rice, ASAE Paper No. 68—809, American Society of Agricultural Engineers, St. Joseph, Mich., 1968.
86. **Chung, D. S. and Converse, H. H.,** Effect of Moisture Content on Some Physical Properties of Grains, ASAE Paper No. 69-811, American Society of Agricultural Engineers, St. Joseph, Mich., 1969.
87. **Osborne, L. E.,** Resistance to airflow of grain and other seeds, *J. Agric. Eng. Res.*, 6, 119, 1961.
88. **Narayan, C. V.,** Behavior of Wheat Grains in Bulk under High Pressures, Master's thesis, University of Guelph, Ontario, Canada, 1966.
89. **Hagues, G. and Russell, J.,** Volumenometer methods. I. Density of barley and malt, *J. Inst. Brewing*, 1969.
90. **Goss, J. R.,** Original data, University of California, Davis, as reported in *Physical Properties of Plant and Animal Materials*, Part II, Vol. Mohsenin, N. M., auth., 1968, 741.
91. **Misener, G. C. and Lee, J. H. A.,** Aerodynamic separation of grain from straw and chaff in a dispersed stream, *Can. Agric. Eng.*, 15, 62, 1973.
92. **Ko, R. S. and Zoerb, G. C.,** Dielectric constant of wheat straw, *Trans. ASAE*, 13, 42, 1970.
93. **Reints, R. E., Jr. and Yoerger, R. R.,** Trajectories of Seeds and Granular Fertilizers, ASAE Paper No. 65-602, American Society of Agricultural Engineers, St. Joseph, Mich., 1965.
94. **Bakker-Arkema, F. W., Rosenau, J. R., and Clifford, W. H.,** Measurements of Grain Surface Area and Its Effect on the Heat and Mass Transfer Rates in Fixed and Moving Beds of Biological Products, ASAE Paper No. 69-356, American Society of Agricultural Engineers, St. Joseph, Mich., 1969.
95. **Mohsenin, N. N.,** Physical properties of plant and animal materials, in *Structure, Physical Characteristics and Rheological Properties*, Vol. 1, Second Preliminary Edition (Part 1), March 1968.
96. **Harmond, J. E., Brandenburg, N. R., and Jensen, L. A.,** Physical Properties of Seed, ASAE Paper No. 63-824, American Society of Agricultural Engineers, St. Joseph, Mich., 1963.
97. **Kazarian, E. A. and Hall, C. W.,** Thermal Properties of Grain, ASAE Paper No. 63-825, American Society of Agricultural Engineers, St. Joseph, Mich., 1963.
98. **Babbitt, E. A.,** The thermal properties of grain in bulk, *Can. J. Res.*, F23, 388, 1945.
99. **Moote, I.,** The effect of moisture on the thermal properties of wheat, *Can. J. Tech.*, 31, 57, 1953.

100. **Sharma, D. K. and Thompson, T. L.,** The Specific Heat and Thermal Conductivity of Grain Sorghum, ASAE Paper No. 72-842, American Society of Agricultural Engineers, St. Joseph, Mich., 1972.
101. **Zink, F. J.,** Specific gravity and air space of grains and seeds, *Agric. Eng.,* 16, 439, 1935.
102. **Browne, D. A.,** Variation of the bulk density of cereals with moisture content, *J. Agric. Eng. Res.,* 7, 288, 1962.
103. Engineering manual of the Industry Machinery Company as reported in *Physical Properties of Plant and Animal Materials,* Vol. 1, (Part 2), Mohsenin, N. M., auth., 1968, 696.
104. **Green, D. E., Cavanah, L. E., and Pinnell, E. L.,** Effect of seed moisture content, field weathering, and combine cylinder speed on soybean seed quality, *Crop Sci.,* 6, 7, 1966.
105. **Gupta, V. K.,** Aerodynamic Grading of Wheat Grain, Master's thesis, University of Guelph, Ontario, Canada, 1968.

HUMAN FACTORS — VIBRATION AND NOISE IN THE WORKPLACE

C. W. Suggs

VIBRATION

Introduction

Many of the machines of modern agriculture produce significant, often intolerable, levels of vibration. Whether or not these high powered, often high speed, machines have intensified the problem is open to speculation. However, it is agreed that vibration levels on many present day field machines fall somewhere between uncomfortable and unsafe.

The trend is toward larger, faster, more powerful machines, which tend to produce higher levels of vibration unless corrective measures are taken. There are, in general, three types of corrective measures that may be taken to protect operators. These are (1) reduction of the vibration at its source, (2) passive vibration isolation between operator and machine, and (3) active vibration feedback control.

Where feasible, vibration should be prevented at its source by attacking the cause. This can be done by better balancing of rotating parts, shock mounting and counterbalancing reciprocating devices, and using softer more resilient tires to reduce vibration inputs from the terrain. Terrain sources account for a large part of the vibration found on farm tractors. This vibration, concentrated around 3 to 6 Hz, is particularly unpleasant.

Vibration Isolation

Passive vibration isolation is traditionally supplied for seats by means of a spring and damper suspension element. On off-road vehicles, the problem is made more complex by the low frequency nature of the vibration against which the seat must be isolated. To be effective, suspensions should be soft enough to provide a natural frequency well below the disturbing frequencies. Such suspensions have excessive static deflections, do not provide operator position control, and must be adjusted to accommodate for different weight operators.

Passive isolation could also be provided by spring mounting of vehicle axles or even wheel rims. However, it has been felt that the resulting lack of vehicle height stability would not provide a good enough reference for control of mounted or perhaps even some trailing tools.

Experimental active vibration isolation equipment has been designed, constructed, and field tested.[1] The equipment involves a vibration sensing device, an electronic signal processing and amplifying section, a source of hydraulic power, an electrohydraulic valve, an actuator (hydraulic cylinder), and a position sensor with a feedback path to the electronic section (Figure 1). In operation, the sensing device detects the onset of motion in the tractor platform, and in conjunction with the rest of the equipment, causes the seat to be driven up as the vehicle frame drops and down as the frame rises. Vibration attenuation in the frequencies between about 1 and 7 Hz was approximately 75%. Vibration in these frequencies, especially the lower end of the range, is particularly difficult to isolate by passive seat suspensions.

There are four parameters of the vibration that must be considered with respect to human responses and tolerances. These are intensity, frequency, direction, and duration. At a given frequency, direction, and duration, vibration tolerance decreases with increasing intensity. Conversely, at a given intensity, tolerance is greater at some frequencies and directions than others. Tolerance always decreases for increased durations of exposure. That is, vibration levels just acceptable for a few minutes would be intolerable for several hours of exposure.

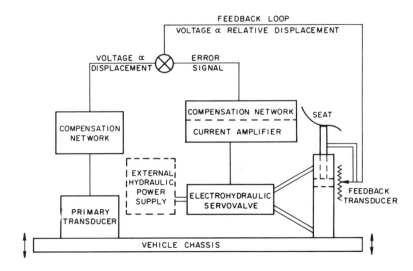

FIGURE 1.   Schematic of the active suspension system.

Mechanical Effects

With respect to dynamic inputs, the human body responds as if it were composed of a series of masses mounted on springs and dampers. The vertical mode has been modeled[2] as a series of six masses with each mass in the sequence supporting those above it through springs and masses. The dominant resonant frequencies of these masses range from about 3 Hz for the chest-abdomen complex to 20 Hz for the head. Many of the components, as well as the seated subject as a unit, resonate at about 5 Hz.[3]

When vibrated at their resonant frequency, the affected body masses may have significant motion with respect to the rest of the body. For example, the viscera possesses a large degree of mobility, which allows for large relative displacements during vibration in the 3 to 5 Hz range. The resulting strain on the mounting structures and on the organs themselves produces discomfort, pain and, if vibration intensity is carried to excess, can rupture an organ or tear it loose from its mounting structure. Resonance of the head can easily be observed during vibration. Resonance of the chest or waist levels, while not so easy to observe, can easily be detected by means of an accelerometer mounted to a belt fitted around the subject at the desired level.[4] It should not be a surprise that man is least tolerant to vibration in the frequency ranges where resonances occur, 2.5 to 10 Hz. The seriousness of the vibration problem for operators of off-road vehicles is accentuated because these vehicles and their seats have their dominant vibrational modes in the same 2.5 to 10 Hz frequency range.[4]

A simplified lumped parameter dynamic model of a seated man containing two spring-damper suspended masses has been developed by the author and associates[3] (Figure 2). The model was not expected to show the dynamics of each body component. It was designed to produce only the overall reaction at the seat level. Since the model was relatively simple, it was possible to evaluate the parameters from a group of subjects and then actually construct the model in hardware (Figure 3a). Parameters were evaluated by driving point mechanical impedance techniques in which the force-velocity relationship necessary to vibrate a subject at a series of frequencies is measured. Parameters are derived from the associated mathematical theory. A comparison of the simulator and subject impedances are given in Figure 3b. The dominant response at about 4.5 Hz is due to the natural frequency of the larger mass.

It has been proposed that such a mechanical model or dynamic simulator be used to represent the average man in dynamic (vibration) tests of vehicle seats. Such an approach would allow seat tests to be standardized so that repeatable results could be expected. Rigid

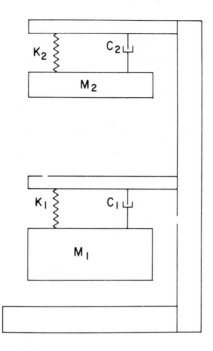

FIGURE 2.   Lumped parameter, two-degree of freedom dynamic analogue of man. (From Suggs, C. W., Abrams, C. F., Jr., and Stikeleather, L. F., *Ergonomics*, 12(1), 79, 1969. With permission.)

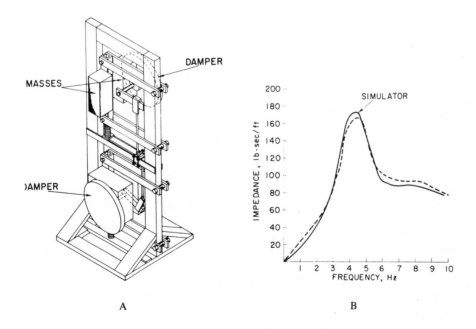

A

B

FIGURE 3.   (A) Dynamic simulator of low frequency human responses. (B) Comparison of dynamic simulator response with average resistance of subjects. (From Suggs, C. W., Abrams, C. F., Jr., and Stikeleather, L. F., *Ergonomics*, 12(1), 79, 1969. With permission.)

mass loading of seats under tests, while allowing for standardization, does not give realistic results because it cannot allow for the resiliency and damping inherent in the human subject. The use of human subjects as seat loads gives meaningful results, but cannot be standardized because of the wide diversity of body size and type as well as the variability of a given subject with time. In addition to overcoming the problems of standardization and realistic representation, the dynamic simulator may be used in long duration studies where human subjects would fatigue and in conditions that might be hazardous to a man's health.

While natural frequencies and masses of the body components have been discussed, some information about damping is needed in order to fully describe the model. Since impedance and transmission curves show resonances, we know that the masses are less than critically damped. In the model just described, the large mass was damped at about 0.25 of critical and the smaller mass at about 0.5 of critical. Other measurements[4] made at the waist, chest, and head levels of vertically vibrated subjects resulted in damping estimates of 0.4, 0.2, and 0.35 of critical, respectively, for the three locations.

For horizontal side-to-side vibration, the seated or standing body has its main resonances between 1 and 3 Hz. The head and shoulders show some resonances at about 1.5 to 2 Hz.[5] For horizontal front-to-rear vibration, resonance has been found in the 2- to 6-Hz range.

Pathological Effects

It has been observed[6] that tractor driving, presumably the associated vibration, has a deleterious effect on the operator's health. Long periods of tractor driving have been shown to produce spinal and stomach disorders. Among a group of tractor operators in the 20- to 30-year age span, 72% were found to have spinal deformations as compared with about 14% for the population as a whole. Surveys carried out in Iowa[7] corroborate this work and conclude that rough tractor rides are at least a contributing cause of many disorders associated with the spine.

Physiological Effects

Oxygen uptake is significantly increased in subjects exposed to vibration intensities above 0.25 g (2.5 m/sec^2) at 2 to 6 Hz in the vertical direction. For the side-to-side and front-to-rear directions, the response was noted at the same acceleration level but at a slightly lower frequency, 1.5 to 4.5 Hz.[4,8]

Since O_2 uptake is directly related to energy expenditure, there is also an increase in energy expenditure with vibration. For farm tractors, this has been shown to be as great as 28% in field experiments.[9] Observations of tractor drivers that show high stress and levels of fatigue confirm these results.

Pulmonary ventilation rate as well as respiration rate appear to be affected by vibration. Ventilation rate increases under exposure to vertical vibration of 0.25 *g* or more in the 2- to 13-Hz frequency range. In many cases, the increase is more than sufficient to supply the additional oxygen needed for the increased O_2 uptake mentioned above. This results in hyperventilation that reduces blood CO_2 levels excessively. Since CO_2 is an acid source, the result is an alkaline condition known as respiration alkalosis. While the condition corrects itself after the vibration is stopped, it can cause drowsiness and loss of attention while the operator is on the machine. This hyperventilation appears to be nonphysiological, but is apparently caused by movements of the organs in the chest and abdomen so as to cause a pumping action in the lungs.

Heart rate increases for vibration in all three directions at intensities of 0.25 to 2.0 *g* at frequencies between 1.5 and 7.5 Hz.[4] This indicates that there is an increase in work load, corroborating the increases in O_2 uptake. Intensities below 0.25 *g* were not great enough to elicit a heart rate response.

Blood pressure increases for vertical vibration at intensities above 0.2 *g* at frequencies of

Table 1
RESPONSES ADVERSELY AFFECTED BY
VIBRATION[11]

Response	Frequency (Hz)	Displacement (in.)
Aiming	15	0.07—0.12
	25	0.035—0.055
	35	0.03—0.05
Hand coordination	2.5—3.5	0.5
Foot pressure control	2.5—3.5	0.5
Hand reaction time	2.5—3.5	0.5
Visual acuity	1—24	0.024—0.59
	35	0.03—0.05
	40	0.065
Tracking	1—5	0.05—0.18
	2.5—3.5	
Attention	2.5—3.5	0.5
	30—3000	0.02—0.20
Hand tremor	20	0.015—0.035
	25	0.035—0.055
	30—300	0.02—0.20

From Woodson, W. E. and Conover, D. W., *Human Engineering Guide for Equipment Designers,* 2nd ed., University of California Press, Berkeley, 1964, chap. 6. With permission.

2.5 to 12 Hz and for frequencies of 4 to 12 Hz in the horizontal directions.[10] Many of the performance responses that are adversely affected by vibration have been summarized by Woodson and Conover,[11] Table 1.

Some work done in our laboratory on lever and pedal tracking showed that performance was decreased by vertical vibration of 0.11 g at 3.7 Hz.[12] Substitution of a spring-damper suspended seat for the rigid seat initially used did not improve performance because the controls moved with the platform instead of the seat. It was felt that mounting the controls on the suspended seat would have regained some of the performance lost due to vibration. In this study, there were no differences in the tracking performance of men and women.

It should be pointed out that a "good" seat is designed to remain at a fixed height while the vehicle frame vibrates beneath it. Thus there may be more relative motion beween the *operator* and the controls than when a rigid seat is used.

Vibration Tolerance Criteria

The acceptability of various levels of vibration is dependent on the criterion of evaluation. Thus levels that are acceptable on the criterion of safety would not be acceptable when based on the criterion of comfort. Carried still further, vibrational levels considered comfortable for a tractor driver might not be acceptable to the same person when a passenger in an aircraft or simply sitting in his own house.

The three following criteria have been proposed by the International Organization for Standardization (ISO):[13] (1) the preservation of comfort, (2) the preservation of working efficiency and performance, and (3) the preservation of health and safety.

From these criteria it is possible to define limits that set the levels of vibrations below which there is little risk the unwanted effect will occur. For example, a comfort limit defines vibrational levels below which exposed subjects would not be expected to experience discomfort.

Three limits have been proposed, corresponding to the three criteria listed above. These are called, respectively, for comfort, working efficiency, and safety: (1) reduced comfort

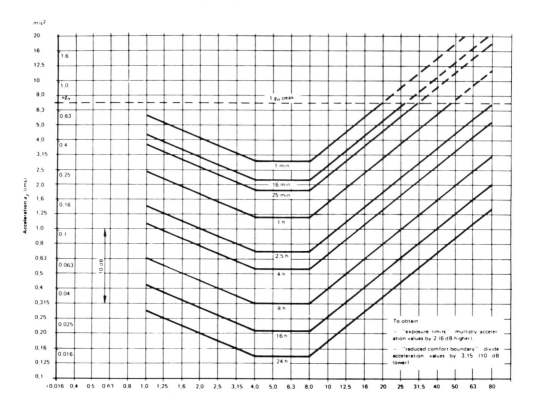

FIGURE 4.   Longitudinal (a_z) acceleration limits as a function of frequency and exposure time; "fatigue-decreased proficiency boundry." (From ISO, International Standard 2631, Guide for the Evaluation of Human Exposure to Whole-Body Vibration, International Organization for Standardization, Geneva, 1969. With permission.)

boundary, (2) fatigue decreased proficiency limit (FDP), and (3) safe exposure limit.

In order to be complete, a limit must be defined for every frequency of vibration in the spectrum of interest. This is most easily done by means of a curve drawn in a frequency-intensity axis. Figures 4 and 5 give the ISO guidelines for fatigue decreased proficiency for various periods of exposure for longitudinal and transverse vibration.[13] Longitudinal vibration is along the length of the body; Figure 6 and transverse vibration is either of the other two perpendicular directions. The Safe Exposure Limit values are twice as great (6 dB higher) as the Fatigue Decreased Proficiency (FDP) limit values. The reduced comfort boundary values may be obtained by dividing the FDP limit values by 3.15 (10 dB lower). The curves in this figure are for one third octave measurements and are applicable to standing or seated subjects. Intensity values should be measured at the body surface not, for example, on a seat surface beneath a cushion. Safety or comfort limits can be developed from Figures 4 and 5 by raising or lowering the FDP limit for the exposure duration desired.

The limit curves for horizontal front to rear or side to side (transverse) vibration exposure are similar in shape to the curves for vertical vibration exposure, but they have been shifted 6 Hz to the left and truncated at 1 Hz. Absence of the curves at low and high frequencies does not mean that there are no limits in these areas, but that insufficient data are available to make recommendations. Vibrational characteristics of farm tractors in various operational situations, wheel loaders, and a lawn tractor are given in Figures 7 and 8.[14] The ISO 8-hr safety and fatigue decreased proficiency boundaries are also drawn in for comparative purposes. These results suggest that good insolation of operator seats from the tractor frame is required in order to reduce operator vibration to acceptable limits.

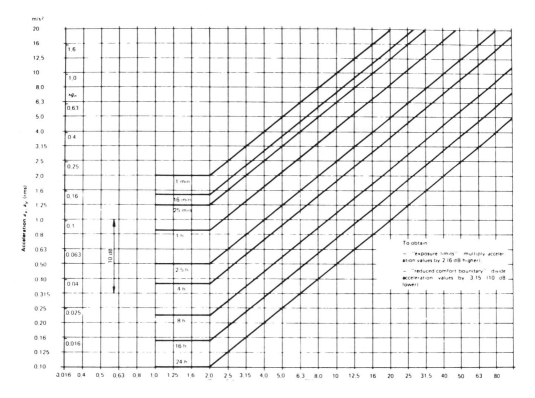

FIGURE 5. Transverse (a_x, a_y) acceleration limits as a function of frequency and exposure time; "fatigue-decreased proficiency boundry." (From ISO, International Standard 2631, Guide for the Evaluation of Human Exposure to Whole-Body Vibration, International Organization for Standardization, Geneva, 1969. With permission.)

Part-Body Vibration

Whereas, off-road vehicles such as tractors, scraper pans, and grain combines exhibit high levels of low frequency (1 to 20 Hz) vibration, they may also have high levels of higher frequency vibration. Because man's sensitivity to higher frequencies is less, they often go unnoticed. Also, the high frequency vibration tends to be isolated by even a poor seat suspension so that the operator is exposed only in the hands and feet from contact with the controls and foot boards.

Hand-held power tools like chain saws produce high levels of high frequency vibration, usually greater than off-road vehicles. The use of such vibrating tools and vehicles has been linked with the occurrence of vibration-induced disorders of the exposed workmen's hands.

There are two general categories of symptoms and signs that have been associated with vibration and found manifested in the hands of workmen who used hand-held power tools. Signs of swelling, skin redness, and cyanosis have been reported among persons using grinders in the airplane industry.[15] Suggs et al.[16] reported hand swelling among operators of chain saws where as much as 4% increase in hand volume of one subject was measured after 45 min of normal operation. The other set of signs and symptoms are those of Raynaud's Disease, which is characterized by a sudden loss of circulation in the hands that results in intense pain. While this disease is known to exist without exposure to vibration, it has been more often reported among workmen who operate hand-held vibrating tools.[17,18] Although there is not complete agreement as to the mechanism of the disease, vascular damage,[19] modification of nerve endings,[20] and mechanical damage to hand muscles[21] taken together provide the most plausible explanation. Abrams[22] gives a comprehensive review of vibration disorders and the conditions felt to be associated with their cause.

Frequency (centre frequency of third octave band) Hz	Weighting factor for			
	longitudinal vibrations (figure 2a)		transverse vibrations (figure 3a)	
1.0	0.50	6 dB	1.00	0 dB
1.25	0.56	5 dB	1.00	0 dB
1.6	0.63	4 dB	1.00	0 dB
2.0	0.71	3 dB	1.00	0 dB
2.5	0.80	2 dB	0.80	2 dB
3.15	0.90	1 dB	0.63	4 dB
4.0	1.00	0 dB	0.5	6 dB
5.0	1.00	0 dB	0.4	8 dB
6.3	1.00	0 dB	0.315	10 dB
8.0	1.00	0 dB	0.25	12 dB
10.0	0.80	2 dB	0.2	14 dB
12.5	0.63	4 dB	0.16	16 dB
16.0	0.50	6 dB	0.125	18 dB
20.0	0.40	8 dB	0.1	20 dB
25.0	0.315	10 dB	0.08	22 dB
31.5	0.25	12 dB	0.063	24 dB
40.0	0.20	14 dB	0.05	26 dB
50.0	0.16	16 dB	0.04	28 dB
63.0	0.125	18 dB	0.0315	30 dB
80.0	0.10	20 dB	0.025	32 dB

1) 4 to 8 Hz in the case of a_z vibration

1 to 2 Hz in the case of a_y or a_x vibration

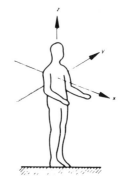
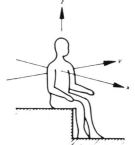
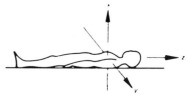

FIGURE 6.   Directions of co-ordinate system for mechanical vibrations influencing humans.[13] (From ISO, International Standard 2631, Guide for the Evaluation of Human Exposure to Whole-Body Vibration, International Organization for Standardization, Geneva, 1969. With permission.)

Critical vibrational frequencies for the production of vibration disorders appear to lie above 100 Hz. Tool speeds of 7500 rpm to 10,000 rpm, (125 to 167 Hz) have been associated with Raynaud's Disease as well as the related signs.[15] It has been observed that chain saw operators, particularly in cold countries, are frequently subject to the disease. Although modern chain saws operate at engine speeds of 100 to 150 Hz (6000 to 9000 rpm) they have been found to produce significant vibration acceleration levels at frequencies well above 3000 Hz.[16] In fact, in some results the highest peaks in the spectra occurred from 1000 to 3000 Hz with significant levels up to 10,000 Hz, while engine operating speed was only 125 Hz, Figure 9.

Transmission of vibration in the arm and hand has been found to depend on frequency.[22] Whereas high frequencies are primarily limited to the hand, the lower frequencies are readily transmitted to the forearm and upper arm (humerus)[23] (Figures 10, 11). The data in the last two figures were obtained by applying vibration to the hand of a cadaver hand-arm system, which was supported at the shoulder by a stand dynamically approximating the shoulder. Acceleration readings were taken from small accelerometers attached directly to the bone just above the wrist and just below and above the elbow. The observed attenuation at high frequencies was more than at low frequencies. Spatial attenuation along the arm was expected

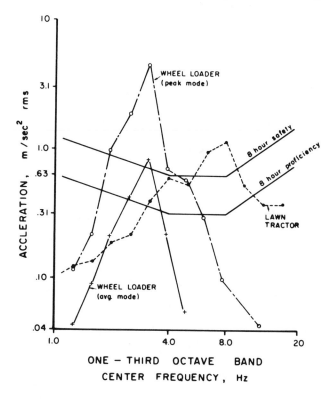

FIGURE 7.   Comparison of the vibrational characteristics of a lawn trac-
tor, a wheel loader, and the 8-hr safety and proficiency limits. (From
Stikeleather, L. F., Hall, G. O., and Radke, A. O., SAE Paper 720001,
Society of Automotive Engineers, New York, 1972. With permission.)

from the spring-mass-damper characteristics of the hand and arm. In fact, the major portion
of the attenuation occurred between the input to the hand and the distal end of the radius
and ulna (the outer ends of the bones in the forearm).

Efforts to model the hand-arm system in order to provide insight with respect to mechanisms and to predict responses have been undertaken by Abrams[24] and Mishoe and Suggs.[25]
They studied the responses of a group of subjects who gripped a handle vibrated sinusoidally
at selected frequencies between 70 and 1500 Hz. They used driving point impedance techniques from which values of damping, stiffness, and dynamic mass could be calculated.
Abrams concluded that for sinusoidal mechanical vibration, the hand can be modeled as a
1° of freedom mass-excited system with a mass of approximately 0.1 lb, a stiffness of
approximately 60 lb_f/in. and a damping factor of approximately 0.6 lb_fsec/in. Thus it appears
that for higher frequencies, only a very small part of the hand is actually vibrating significantly. It would be logical to assume that the surface of the palm is the part involved.

Vibration acceleration limits for hand-held tools have been suggested to the ISO.[26] Figure
12 gives the coordinate system for measuring and reporting hand vibration, and Figure 13
gives the allowable one third octave band levels for hand-arm vibration in any of the three
coordinate directions.[26] Figure 13 contains lines for correction factors 1 through 5, which
are related to duration of exposure according to Table 2. If vibration measurements are made
in full octave rather than one third octaves, Figure 14 can be used to determine the appropriate
guidelines. The logic of the ISO exposure guidelines is illustrated by noting that an equal
sensation curve of acceleration vs. frequency has approximately the same slope as the ISO
exposure guideline (Figure 15).

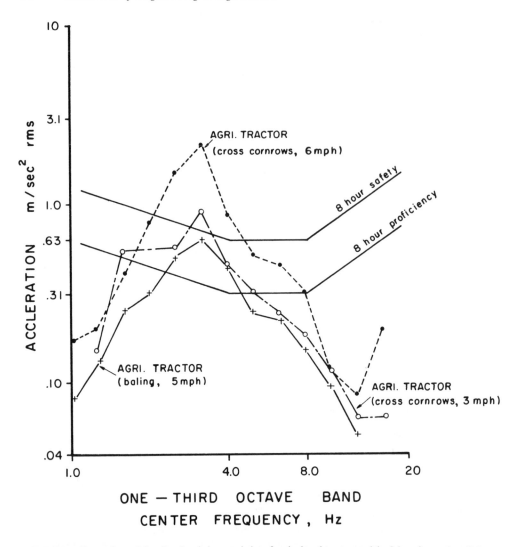

FIGURE 8.   Comparison of the vibrational characteristics of agricultural tractors and the 8-hr safety and proficiency limits. (From Stikeleather, L. F., Hall, G. O., and Radke, A. O., SAE Paper 720001, Society of Automotive Engineers, New York, 1972. With permission.)

Many machines in common usage violate the hand-arm guideline for 8 hr of exposure, and some violate the 0.5 hr exposure guideline (Figures 16 to 21). Fortunately, many of these tools are used only part-time or intermittently, for example, by a carpenter who would spend much of his time nailing and fitting pieces of lumber into place.

NOISE

Introduction

Any unwanted sound is, by definition, noise in the ear of the observer. It follows from this definition that sound desired by one individual may be noise to another. For our purposes here, we will also consider that sound levels high enough to damage the ear are noise whether desired or not.

Sound is airborne vibration and as such, has the properties of frequency and intensity associated with vibration. However, since sound may be propagated in all directions from

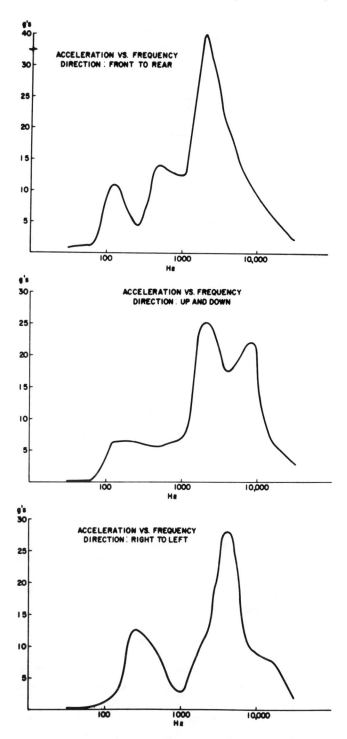

FIGURE 9.   Typical vibration characteristics of chain-saw handles without vibration isolators. (From Suggs, C. W., Abrams, C. F., Jr., and Cundiff, J. S., *J. Sound Vib.*, 2(6), 18, 1968. With permission.)

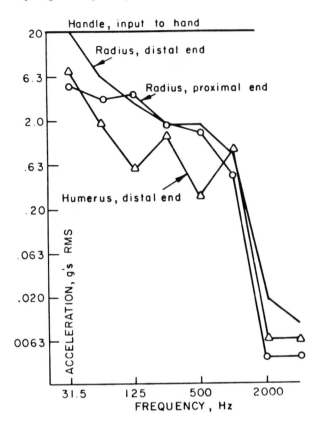

FIGURE 10.   Vibration attenuation in hand and arm. (From Suggs, C. W., 4th Int. Ergonomics Conf., Strasbourg, France, July, 1970. With permission.)

a source, the concept of direction and phase between components has less meaning in sound than in vibration.

While noise is known to distract, confuse, and make mental concentration more difficult, the main effect is hearing loss. Noise-induced hearing loss does not usually occur in a rapid traumatic manner. In fact, the problem may be more acute and widespread because the loss occurs imperceptibly, slowly, and without pain. Noises loud enough to produce permanent hearing loss often produce a temporary threshold shift and ringing in the ears. Thus, if a workman finds that his ears ring or that he cannot hear well after several hours of exposure to a given noise, it is likely that some permanent hearing loss is taking place.

Depending on the individual, the noise intensity, and the length of daily or weekly exposure, it may take several years for significant hearing loss to occur. Losses are usually characterized by a decrease in hearing acuity at 4000 Hz followed by a spread into the conversational (lower) frequencies. While hearing losses at 4000 Hz may decrease ones enjoyment of music, losses in the conversational frequencies seriously affect ones ability to communicate. It is generally conceded that 40 hr of exposure per week to noises of 90 dBA or greater will result in hearing loss. The letter "A" refers to a frequency weighting scale (Figure 22) which is shaped approximately like human hearing acuity.[27] This is the level given by the federal Walsh-Healy Act as being the maximum allowable 8 hr/day exposure unless ear muffs or plugs are worn by the employee. The conditions of the Walsh-Healy Act are given below:[28]

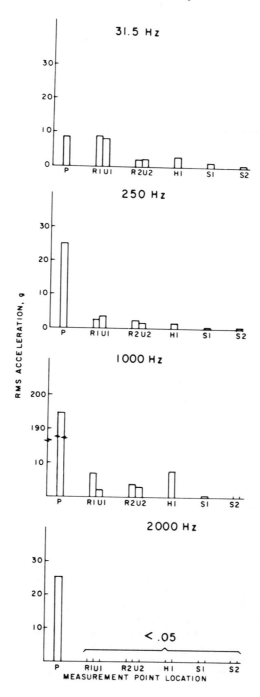

FIGURE 11.   Spatial array of vibration levels at several points along the arm. (From Abrams, C. F., Jr., Master's thesis, North Carolina State University, Raleigh, 1969. With permission.)

"Handgrip" position

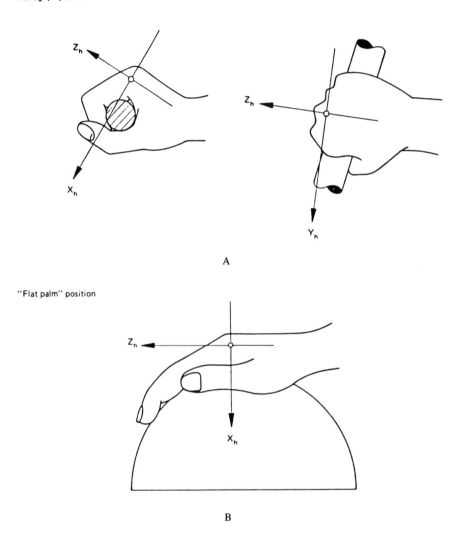

A

"Flat palm" position

B

FIGURE 12.   Coordinate system for the hand. (A) In the "handgrip" position (standardized grip on a cylindrical bar of radius 2 cm). (B) In the "flat-palm" position (hand pressed onto ball of radius 5 cm). Note: (The origin of the system is deemed to lie in the head of the third metacarpal and the Z-axis to be defined by the longitudinal axis of that bone. The X-axis is projecting anteriorly from the origin when the hand is in the normal anatomical position — palm facing forwards. The Y-axis passes through the origin and is perpendicular to the X-axis.) (From ISO, Guide for the Measurement and the Evaluation of Human Exposure to Vibration Transmitted to the Hand, Draft Proposal No. 5349, International Organization for Standardization, Geneva, 1978. With permission.)

Walsh-Healey Public Contracts Act
As reprinted from Safety and Health Standards, Federal Register, 35, 7946-7954 (May 20, 1969).
(As revised by errata sheet dated July 15, 1969)

Rules and Regulations

50-204.10 Occupational noise exposure

   (a) Protection against the effects of noise exposure shall be provided when the sound levels exceed those shown in Table 1 of this section when measured on the A scale of a standard sound level meter at slow response. When noise levels are determined by octave band analysis, the equivalent A-weighted sound level may be determined as follows:

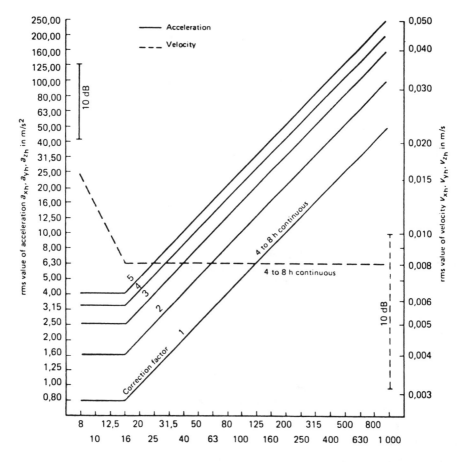

FIGURE 13. Exposure guidelines for hand-arm vibration. (From ISO, Guide for the Measurement and the Evaluation of Human Exposure to Vibration Transmitted to the Hand, Draft Proposal No. 5349, International Organization for Standardization, Geneva, 1978. With permission.)

(b) When employees are subjected to sound levels exceeding those listed in Table 1 of this section, feasible administrative or engineering controls shall be utilized. If such controls fail to reduce sound levels within the levels of the table, personal protective equipment shall be provided and used to reduce sound levels within the levels of the table.

(c) If the variations in noise level involve maxima at intervals of 1 sec or less, it is to be considered continuous.

(d) In all cases where the sound levels exceed the values shown herein, a continuing, effective hearing conservation program shall be administered.

Exposure to impulsive or impact noise should not exceed 140 dB peak sound pressure level.

The human ear has a very wide dynamic range, that is, it can sense very low levels as well as very high levels without being overdriven. Intensities up to 120 dB can be sensed without pain. Since 20 dB is equal to $10^1:1$ change in pressure, the entire range from 0 to 120 dB is equal to a $10^6:1$ pressure change, that is, a million to one. If the comparison is made on a power scale instead of on the basis of pressure, the dynamic range is $10^{12}:1$ because pressure is proportional to power squared.

The reference pressure of 0.0002 μbar/cm² is equivalent to a power of about 10^{-16} W/cm². When the area of the ear canal is considered (less than 1 cm²), it is seen that the ear is extremely sensitive. In fact, in absolute sensitivity it compares favorably with well designed, finely tuned electronic equipment. For example, a microvoltmeter would have to draw no more than 10^{-10} A to equal the sensitivity of the ear.

Table 2
PERMISSIBLE NOISE
EXPOSURES[a]

Duration/per day (hr)	Sound level dBA, slow response
8	90
6	92
4	95
3	97
2	100
1 1/2	102
1	105
1/2	110
1/4 or less	115

[a] When the daily noise exposure is composed of two or more periods of noise exposure of different levels, their combined effect should be considered, rather than the individual effect of each. If the sum of the following fraction: C1/T1 + C2/T2...Cn/Tn exceeds unity, then the mixed exposure should be considered to exceed the limit value. Cn indicates the total time of exposure at a specified noise level, and Tn indicates the total time of exposure permitted at that level.

Instrumentation

Sound level meters are supplied by several manufacturers. They can be fitted with an octave band filter set so that the frequency characteristics of the sound source may be determined directly. The unit is supplied with a meter that can be read as well as with an electrical output that can be used as an input to a recorder. The microphone is attached directly to the instrument, but may be placed on an extension cable for remote sensing.

Noise may be measured with a condenser type microphone consisting of a thin diaphragm, which is caused to deflect by a sound pressure field. The motion is detected by electronic means, producing an electrical signal proportional to the sound level. After amplification, the signal may be displayed on a meter, recorded on magnetic tape, or on various other display means.

Farm Machinery Noise

Noise, always a problem on most farm machines, has been successfully attenuated in many cases. However, the problem still exists, especially on small engine-powered devices where space, weight, and power limitations make installation of adequate muffling equipment difficult.

Typical farm machines produce noise levels in excess of 100 dB, and machines without mufflers have been observed to produce levels as high as 120 dB on a linear unweighted scale (Figure 23). In some of the older data, cabs did not reduce noise exposures, but this has been corrected in more recent years. Spectrograms of the noise from these sources usually are highest in the 63 to 250 Hz range (Figures 23 and 24). Chain saws, with their higher-rpm engines tend to have maximum intensity at higher frequencies (Figure 25).

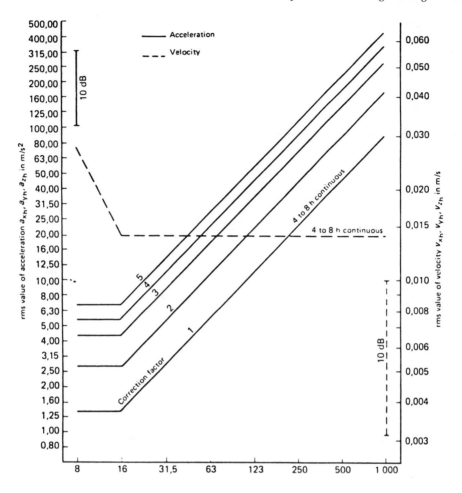

FIGURE 14.   Exposure guidelines for hand-arm vibration. (From International Organization for Standardization, Guide for the Measurement and the Evaluation of Human Exposure to Vibration Transmitted to the Hand, Draft Proposal No. 5349, 1978. With permission.)

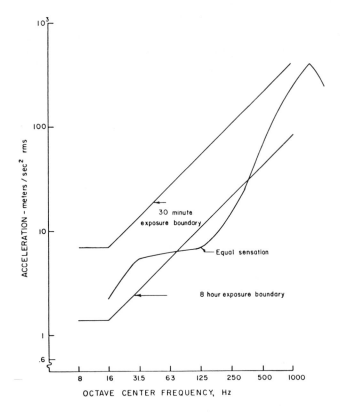

FIGURE 15. Comparison of the ISO 8-hr exposure boundary with an equal sensation line. (From Mishoe, J. W., and Suggs, C. W., *J. Sound Vib.*, 54, 545, 1977. With permission.)

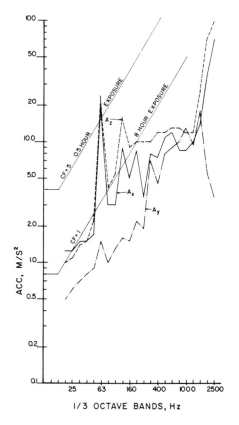

FIGURE 16. One third octave acceleration spectrogram of a reciprocating electric sander. Three directions of vibration, ISO exposure guidelines superimposed.

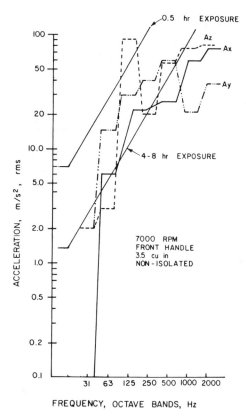

FIGURE 17. One third octave acceleration spectrogram of a gasoline powered chain saw without vibration isolation. Three directions of vibration, ISO guidelines superimposed.

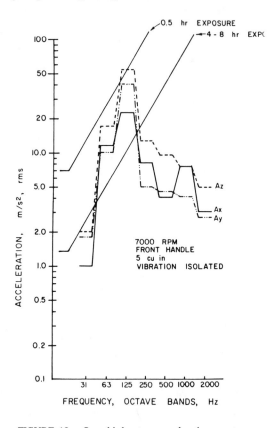

FIGURE 18.   One third octave acceleration spectrogram
of gasoline powered chain saw from Figure 17 with vibration
isolated handles. Three directions of vibration, ISO exposure guidelines superimposed.

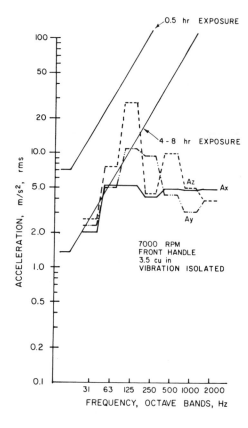

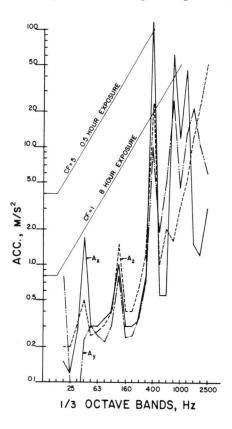

FIGURE 19.   One third octave acceleration spectrogram of a larger, gasoline powered, chain saw with vibration isolated handles. Three directions of vibration, ISO exposure guidelines superimposed.

FIGURE 20.   One third octave acceleration spectrogram of an electric powered router operating on dry pine. Three directions of vibration, ISO exposure guidelines are superimposed.

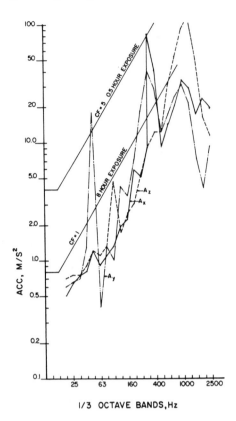

FIGURE 21.   One third octave acceleration spectrogram of an electric powered reciprocating saber saw cutting dry pine. Three directions of vibration, ISO guidelines superimposed.

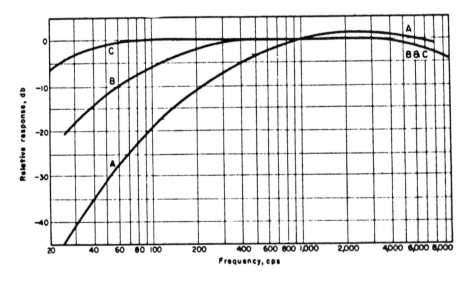

FIGURE 22.   Weighting curves for sound measurements.

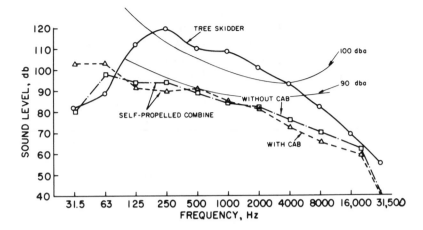

FIGURE 23.   Octave band spectrograms of a tree skidder and a self-propelled combine in comparison to the 90 and 100 dB (A) contours.

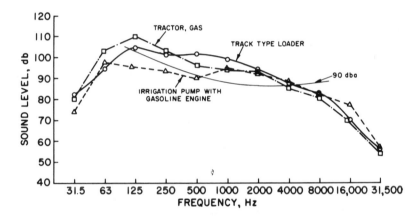

FIGURE 24.   Octave band spectrograms of a tractor, loader, and gasoline powered irrigation pump in comparison to the 90 dB(A) contour.

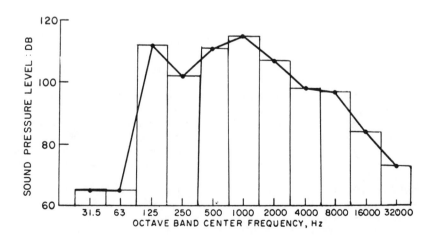

FIGURE 25.   Octave band analysis of chain saw noise. (Author's unpublished data).

Table 3
WEIGHTING FACTORS RELATIVE TO
THE FREQUENCY RANGE OF MAXIMUM
ACCELERATION SENSITIVY[a] FOR THE
RESPONSE CURVES OF FIGURES 2a AND
3a

Frequency (center frequency of third-octave band) Hz	Weighting factor for	
	Longitudinal vibrations (dB) (Figure 2a)	Transverse vibrations (dB) (Figure 3a)
1.0	0.50 = −6	1.00 = 0
1.25	0.56 = −5	1.00 = 0
1.6	0.63 = −4	1.00 = 0
2.0	0.71 = −3	1.00 = 0
2.5	0.80 = −2	0.80 = −2
3.15	0.90 = −1	0.63 = −4
4.0	1.00 = 0	0.5 = −6
5.0	1.00 = 0	0.4 = −8
6.3	1.00 = 0	0.315 = −10
8.0	1.00 = 0	0.25 = −12
10.0	0.80 = −2	0.2 = −14
12.5	0.63 = −4	0.16 = −16
16.0	0.50 = −6	0.125 = −18
20.0	0.40 = −8	0.1 = −20
25.0	0.315 = −10	0.08 = −22
31.5	0.25 = −12	0.063 = −24
40.0	0.20 = −14	0.05 = −26
50.0	0.16 = −16	0.04 = −28
63.0	0.125 = −18	0.0315 = −30
80.0	0.10 = −20	0.025 = −32

[a]   4 to 8 Hz in the case of $\pm a_z$ vibration.
     1 to 2 Hz in the case of $\pm a_y$ or $\pm a_x$ vibration.

REFERENCES

1. **Stikeleather, L. F. and Suggs, C. W.,** An active seat suspension system for off-road vehicles, *Trans. ASAE,* 13(1), 99, 1970.
2. **Bryce, W. D.,** A Review and Assessment of Criteria for Human Comfort Derived from Subjective Responses to Vibration, Ministry of Tech. National Gas Turbine Establishment, Pyestock, Hants, Dec., 1966.
3. **Suggs, C. W., Abrams, C. F., Jr., and Stikeleather, L. F.,** Application of a damped spring-mass human vibration simulator in vibration testing of vehicle seats, *Ergonomics,* 12(1), 79, 1969.
4. **Suggs, C. W. and Huang, B. K.,** Seat Suspension Characteristics and Operator Response to Farm Machinery Vibration, 13th Int. Congr. Sci. Work-Organization Agric., Brussels, Rijksstation voor Boerder lijbouwkunde, June 1966.
5. **Dieckmann, D.,** Influence of horizontal mechanical vibration on man, *Int. Z. Angew. Physiologie Einschl. Arbeitsphysiologie,* 17, 83, 1958.
6. **Roseggar, R. and Roseggar, S.,** Health effects of tractor driving, *J. Agric. Eng. Res.,* 5(2), 241, 1960.
7. **Fishbein, W. and Salter, L. C.,** The relationships between truck and tractor driving and disorders of the spine and supporting structures, *Ind. Med. Sur.,* 19, 1950, 444.
8. **Hornick, R. J. and Lefrits, N. M.,** A study and review of human response to prolonged random vibration, *Human Factors,* 8, 481, 1966.

9. **Dupuis, H. R.,** Farm Tractor Operation and Human Stress, Paper presented at ASAE Annual Meeting, Chicago, Ill., December 1957.

10. **Mozell, M. M.,** Behavioral effects of whole-body vibration, *J. Aviation Med.,* 29, 716, 1958.

11. **Woodson, W. E. and Conover, D. W.,** *Human Engineering Guide for Equipment Designers,* 2nd ed., University of California Press, Berkeley, 1964, chap. 6.

12. **Hansson, J. E. and Suggs, C. W.,** 1973. Institutionen for Skogstecknik Skogshagiskolon, Stockholm, Nr. 63. Lagfrekventa vibrationers effekt pa reglagemanovrering (The effect of seat vibration on vehicle operators' lever and pedal control capabilities).

13. ISO, International Standard 2631, Guide for the Evaluation of Human Exposure to Whole-Body Vibration, International Organization for Standardization, Geneva, 1969.

14. **Stikeleather, L. F., Hall, G. O., and Radke, A. O.,** A Study of Vehicle Vibration Spectra as Related to Seating Dynamics, SAE Paper 720001, Society of Automotive Engineers, New York, 1972.

15. **Dart, E. E.,** Effects of high speed vibrating tools on operators engaged in the airplane industry, *Occup. Med.,* 1(6), 515, 1946.

16. **Suggs, C. W., Abrams, C. F., Jr., and Cundiff, J. S.,** Attenuation of high frequency vibration in chain saws, *J. Sound Vib.,* 2(6), 18, 1968.

17. **Agate, J. N. and Druett, H. A.,** A study of portable vibrating tools in relation to the clinical effects which they produce, *Br. J. Ind. Med.,* 4, 141, 1947.

18. **Ashe, W. F.,** Physiological and Pathological Effects of Mechanical Vibration on Animals and Man, Rep. 862-5, The Ohio State University Research Foundation, Columbus, 1964.

19. **Telford, E. D., McCann, M. B., and MacCormack, D. H.,** 1945 "Dead Hands" in users of vibratory tools, *Lancet,* 1, 359, 1945.

20. **Takats de, G. and Fowler, E. F.,** Raynaud's phenomenon, *JAMA,* 179(1), 99, 1962.

21. **Jepson, R. P.,** Raynaud's phenomenon in workers with vibrating tools, *Br. J. Ind. Med.,* 11, 183, 1954.

22. **Abrams, C. F. Jr.,** A Study of the Transmission High Frequency Vibration in the Human Arm, Master's thesis, North Carolina State University, Raleigh, 1969.

23. **Suggs, C. W.,** Vibration of Machine Handles and Controls and Propagation through the Hands and Arms, 4th Int. Ergonomics Congr., Strasbourg, France, July 1970.

24. **Abrams, C. F., Jr.,** Modeling the Vibrational Characteristics of the Human Hand by the Driving Point Mechanical Impedance Method, Ph.D. thesis, North Carolina State University, Raleigh, 1971.

25. **Mishoe, J. W. and Suggs, C. W.,** Hand arm vibration. II. Vibrational responses of the human hand, *J. Sound Vib.,* 54, 545, 1977.

26. ISO, Guide for the Measurement and the Evaluation of Human Exposure to Vibration Transmitted to the Hand, Draft proposal No. 5349, International Organization for Standardization, Geneva, 1978.

27. **Hirschum, M.,** The 90 dBA guideline for noise control engineering, *J. Sound Vib.,* 4, 25, 1970.

28. **Federal Register,** Safety and Health Standards Walsh-Healy Public Contracts Act, July 15, 1969.

HUMAN FACTORS — THERMAL COMFORT, WORKLOADS, FORCES AND ANTHROPOMETRICS

C. W. Suggs

THERMAL COMFORT, CLOTHING, AND WORKLOADS

Thermal comfort is determined primarily by skin temperature. Skin temperature is determined in turn by environmental factors, workload, and the insulating qualities of the clothing being worn. A skin temperature of about 93°F is sensed as comfortable while a person with a skin temperature of 98°F is very hot, and one with a skin temperature of 84°F is very cold[1] (Table 1). The temperature and humidity ranges that will produce a comfortable skin temperature are shown in Figure 1. Lower temperatures can be tolerated in the hands and feet more than in the rest of the body (Table 2).

When temperatures above the optimum range must be tolerated, it may be necessary to limit the duration of exposure (Figure 2). While some protective clothing is available for high temperatures, it is commonly used to protect against low temperatures. Figure 3 gives the relationship between allowable exposure duration and temperature for four kinds of clothing from light to heavy.[1] The heavy clothing, which had four times the insulating value of the lightest, would protect a person against 0°C weather for days, whereas the lightest clothing would provide only about 1 hr of tolerance time.

The insulating quality is rated in clo units, which is the insulating value of a typical business suit equal to 0.18°C m² hr/kcal.[2] Table 3 gives the clo values of various clothing ensembles. An active man needs less clothing in a cold environment than an inactive one. For example, a man sleeping outdoors at −40°C needs clothing with an insulating value of about 12 clo units in order to maintain thermal balance. When the same man is physically active (walking, for example) he will need only one third as much insulation in his clothes, or four clo units, because he will be producing about three times as much heat as when sleeping.[3]

In addition to amount of clothing, comfort is dependent on air temperature, air velocity, humidity or wet bulb temperature, and workload (Figure 4).[2] Increases in mean radiant temperature can also significantly decrease the comfortable dry bulb temperature (Figure 5).[2] The relationship between convective heat loss and evaporative heat loss allows the total energy absorbing potential of the air to be expressed in terms of enthalpy.[4] The dry bulb equivalence of thermal radiation has been derived and evaluated to determine the amount thermal radiation may increase to compensate for a 1° decrease in air temperature (Figure 6).[2,5]

Workloads appreciably affect the environmental conditions at which people will be most comfortable. Workloads or energy expenditures vary from less than 1 kcal/min for light work to 20 or more for heavy work (Table 4).[6] In general, females expend less energy per day than men in similar occupations (Table 5).[6] Figure 7 shows how much of the total energy consumption for various tasks appears as work and how much as heat,[6] and Table 6 shows the energy expenditure rate for various activities.

FORCE CAPABILITIES AND ANTHROPOMETRIC MEASUREMENTS

Lifting ability varies appreciably with lift height and with the size of the subject (Table 7).[8] Hand grip strength averaged approximately 100 lb for men, being less for the left hand, smaller individuals, and women (Table 8).[8]

The force which can be exerted against a pedal by the leg is dependent on the location of the pedal with respect to the seat reference point (SRP) (Figure 8).[9] Large forces cannot

Table 1
RELATION BETWEEN
COMFORT AND MEAN
SKIN TEMPERATURES (°F)

Very hot	98
Unpleasantly warm	96
Indifferently warm	94
Comfortable	93
Comfortably cool	91
Indifferently cool	88
Unpleasantly cool	86
Extremely cold	84

From Woodson, W. W. and Conover,
D. W., *Human Engineering Guide for
Equipment Designers,* 2nd ed., University of California Press, Berkeley,
1964, chap. 6. With permission.

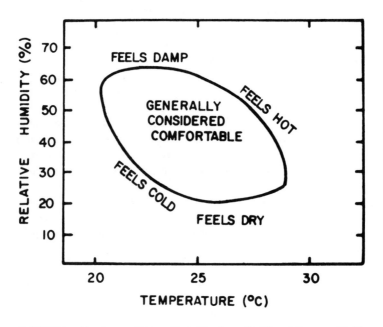

FIGURE 1.   Comfort conditions. (From Woodson, W. W. and Conover, D. W.,
Human Engineering Guide for Equipment Designers, 2nd ed., University of California Press, Berkeley, 1964, chap. 6. With permission.)

be sustained for a very long time.[8] Table 9 shows that doubling the force decreases the time
which the force can be held by a factor of four or five.

Position of the upper arm has a significant effect on the force that can be exerted by the
arm (Figure 9), and push-pull strength is greater than lift-lower strength. Side-to-side strength
tends to be less than either of the above, due at least partly to the absence of a reaction
support to brace against in most working situations.

That maximum strength of the hand and leg varies with time has already been mentioned.
Generally the force that can be held for an extended length of time is about 25% of maximum
strength, Figure 10.[8] Because of the limits imposed by age, body size, sex, and the decrease
in motion speed with increased loads, tasks and workplaces should generally be designed

Table 2
RELATION BETWEEN
COMFORT AND
TEMPERATURE OF THE
EXTREMITIES (°F)

	Hands	Feet
Minimum	68	73
Tolerable	68—59	73—64
Intolerable pain	59—50	64—55
Numbness	50—	55—

From Woodson, W. W. and Conover,
D. W., *Human Engineering Guide for
Equipment Designers,* 2nd ed., University
of California Press, Berkeley, 1964, chap.
6. With permission.

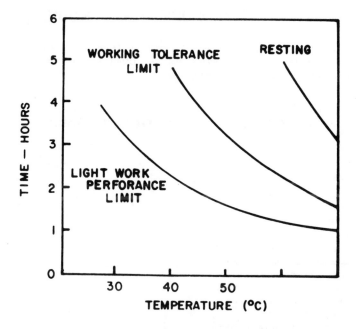

FIGURE 2.   Tolerance to heat.[2] (From Fanger, P. O., *Thermal Comfort
Analysis and Applications in Environmental Engineering,* Danish Technical
Press, Copenhagen, 1970, chap. 2. With permission.)

to require no more than about 25% of one's maximum strength. If a load or device needs
to be moved rapidly, the force should be reduced to a point well below 25% of maximum
strength.

Leg reach for the seated subject varies from about 20 to 40 in. depending on position and
whether the toe or heel is involved (Figure 11).[9] While the leg and foot are capable of
precision control, the foot does not normally have grasp capability. Therefore, the hand is
more versatile. Figures 12 and 13 give arm reach for the fifth percentile man in vertical and
horizontal planes located at various distances from the seat reference point. Values for Figure
13 are given in Table 10.

Vision is often seriously limited by the obstructions in the workplace, for example by

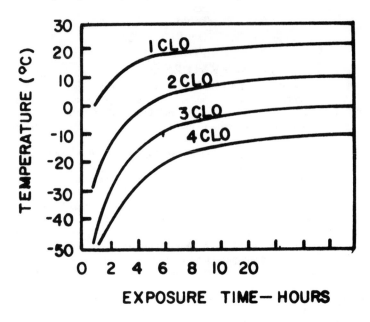

FIGURE 3.   Tolerance to cold. (From Fanger, P. O., *Thermal Comfort Analysis and Applications in Environmental Engineering,* Danish Technical Press, Copenhagen, 1970, chap. 2. With permission.)

Table 3
DATA FOR DIFFERENT CLOTHING ENSEMBLES

Clothing ensemble	I_{cl} clo	f_{cl}^{a} Nude area/ clothed area
Nude	0	1.0
Shorts	0.1	1.0
Typical tropical clothing	0.3—0.4	1.5
Shorts, open-neck shirt with short sleeves, light socks, and sandals		
Light summer clothing	0.5	1.1
Long light-weight-trousers, open neck shirt with short sleeves		
U.S. Army "Fatigues"	0.7	1.1
Light-weight underwear, cotton shirt with trousers, cushion sole socks and combat boots		
Tropical combat uniform	0.8	1.1
Same general components as U.S. Army fatigues but with shirt and trousers of wind resistant poplin		
Typical business suit	1.0	1.15
Heavy traditional European business suit, with long cotton underwear	1.5	1.15—1.2
U.S. Army standard cold-wet uniform	1.5—2.0	1.3—1.4
Cotton-wool underwear, wool and nylon flannel shirt, wind resistant, water repellant trousers and field coat, cloth mohair and wool coat liner and wool socks		
Polar weather suit	3—4	1.3—1.5
Heavy wool pile		

Table 3 (continued)
DATA FOR DIFFERENT CLOTHING ENSEMBLES

Clothing ensemble	I_{cl} clo	f_{cl}[a] Nude area/ clothed area
Eskimo suit[b] Two layers of Caribou fur, $1\frac{1}{2}$ in. thick per layer	12	

[a]   Clothed body is slightly larger.
[b]   See Reference 3.

From Fanger, P. O., *Thermal Comfort Analysis and Applications in Environmental Engineering*, Danish Technical Press, Copenhagen, 1970, chap. 2. With permission.

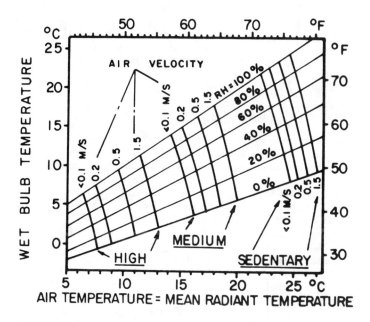

FIGURE 4.   The approximately vertical lines describe equal comfort conditions for person wearing medium clothing (I_{cl} = 1.0 clo, f_{cl} = 1.15) at high, medium, and sedentary work levels.

dashboards, cab cornerposts, hood lines, etc. Figure 14 gives the location of the eye for various lines of sight.[8]

Anthropometric measurements of the male and female body are given in Table 11 and located on the body in Figure 15.

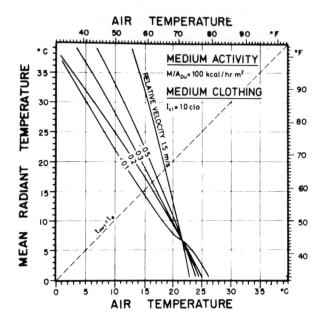

FIGURE 5.   Effects of mean radiant temperature on comfort; air velocity as a parameter. (From Fanger, P. O., *Thermal Comfort Analysis and Applications in Environmental Engineering,* Danish Technical Press, Copenhagen, 1970, chap. 2. With permission.)

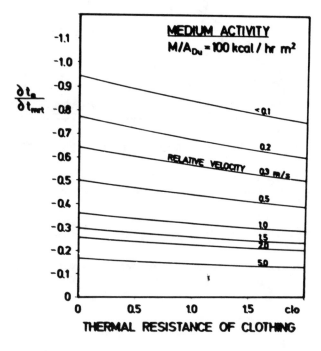

FIGURE 6.   The change in air temperature required to maintain optimal comfort when the mean radiant temperature is increased 1°C, dt_a/dt_{mrt}, vs. the thermal insulation resistance of the clothing in Clo units.[2,5]

Table 4
ENERGY CONSUMPTION IN WORK CALORIES (TOTAL METABOLISM DURING WORK MINUS BASAL METABOLISM)[6,7]

Activity	Conditions of work	kcal/min
Walking without load	Level, smooth surface (4 km/hr)	2.1
	Ploughed soil (3 km/hr)	5.2
	Country road, heavy shoes (4 km/hr)	3.1
Walking, load on back	Level, firm surface	
	10 kg load (4 km/hr)	3.6
	30 kg load (4 km/hr)	5.3
	50 kg load (4 km/hr)	8.1
Climbing	16° Gradient, 11.5 m/min	8.3
	without load	
	20 kg load	10.5
	50 kg load	16.0
Climbing stairs	30.5° Gradient, 17.2 m/min	13.7
	without load	
	20 kg load	18.4
	50 kg load	26.3
Cycling	Speed 16 km/hr	5.2
Pulling cart	3.6 km/hr, level firm surface, pulling energy	8.5
	11.6 kg	
Working with axe	Two-handed strokes, 35 strokes/min	9.5—11.5
Working with hammer	4.4 kg Weight of hammer, 15 strokes/min	7.3
Working with shovel	Throwing distance 2 m	
	Throwing height 1.0 m (10 strokes/min)	7.8
	Throwing height 1.5 m (10 strokes/min)	9.0
	Throwing height 2.0 m (10 strokes/min)	10.0
Sawing wood	Two-man cross-cut saw, 60 double pulls/min	9.0
Bricklaying	Normal output 0.0041 m³/min	3.0
Digging	Garden spade in clay soil	7.5—8.7
Mowing	Scythe	8.3
Milking	180 Pulls/min	2.2
Household chores	Cooking	1—2
	Light cleaning or ironing	2—3
	Washing floors	4—5

Table 5
ENERGY EXPENDITURE IN DIFFERENT OCCUPATIONS[6,7]

Men kcal/day	Women kcal/day	Type of work	Occupations
2400	2000	Sedentary, light manual work	Bookkeeper
2700	2250	Sedentary, light manual work	Shorthand typist
			Watchmaker
		Standing, light manual work	Hairdresser
		Walking	Shepherd (flat country)
3000	2500	Sedentary, heavy manual work	Weaver
			Basket worker
		Sedentary, heavy work involving arms	Bus driver
		Standing, light work involving whole body	Mechanical engineer
		Walking light manual work	Fitter (mech.)
			General practitioner
		Climbing stairs	Meter reader
3300	2750	Sedentary, heavy manual work	Shoemaker

Table 5 (continued)
ENERGY EXPENDITURE IN DIFFERENT OCCUPATIONS[6,7]

Men kcal/day	Women kcal/day	Type of work	Occupations
		Standing, heavy work involving arms	Engine driver
		Walking, light body work	Electrical fitter
		Stair climbing, light body work	Postman
3600	3000	Sitting, heavy work involving arms	Stonemason working on pavings
		Standing, medium body work	Locksmith (fitter)
		Walking, medium body work	Housewife, butcher
		Climbing, heavy work involving arms	Steeplejack
3900	3250	Standing, very heavy body work	Sawing firewood
		Climbing, medium heavy body work	Carpenter on a building site
4200	—	Standing, extremely heavy body work	Coal miner
		Walking, very heavy body work	Agricultural labourer
		Climbing, heavy body work	Worker in a vineyard
4500	—	Standing, extremely heavy body work	Lumberjack
		Walking, very heavy body work	Coal cutter
4800	—	Awkward position, extremely heavy body work	Coal hewer
5100	—	Walking, extremely heavy body work	Harvesting

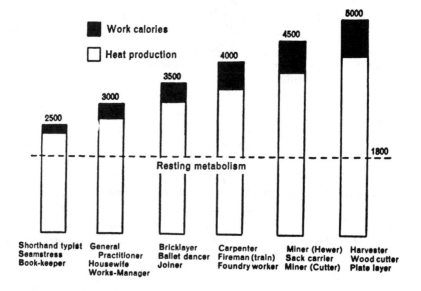

FIGURE 7.   Heat production in different occupations. The height of the columns and the figures give the total calorie consumption per 24 hr. The black parts represent the work calories and the white parts the heat production.[6,7]

Table 6
ENERGY EXPENDITURE FOR
VARIOUS ACTIVITIES

Sleep	60 to 70 kcal/hr
Light work (sitting)	100 to 115 kcal/hr
Flying aircraft	100 to 200 kcal/hr
Light but vigorous exercise	350 to 500 kcal/hr
Heavy, vigorous exercise	500 to 750 kcal/hr

From Woodson, W. W. and Conover, D. W., *Human Engineering Guide for Equipment Designers,* 2nd ed., University of California Press, Berkeley, 1964, chap. 6. With permission.

Table 7
LIFTING ABILITY OF YOUNG FIFTH PERCENTILE MEN

Height lifted from floor (ft)	Maximum weight lifted by 5th percentile men (lb)	95 percentile
1	142	301
2	139	259
3	77	172
4	55	112
5	36	83

From Damon, H., Stout, H. W., and McFarland, R., *The Human Body in Equipment Design,* Harvard University Press, Cambridge, 1966, chap. 2. With permission.

Table 8
HAND STRENGTH: DYNAMOMETER SQUEEZE

	Maximum squeeze — pounds percentile		
	Mean	5th	95th
Men			
Military personnel			
Right hand	119	82	156
Left hand	109	76	144
Industrial workers and bus and truck drivers			
Preferred hand	119	92	146
General population			
Right hand	108	74	142
Left hand	95	65	124
General population, maximum strength held for 1 min.			
Right hand	62	42	82
Left hand	55	39	71

Table 8 (continued)
HAND STRENGTH: DYNAMOMETER
SQUEEZE

	Maximum squeeze — pounds percentile		
	Mean	5th	95th
Women			
Military Personnel, mean of 2 hands	73	58	87
Industrial workers preferred hand	62	45	80
Students preferred hand	63	43	82

From Damon, H., Stout, H. W., and McFarland, R., *The Human Body in Equipment Design,* Harvard University Press, Cambridge, 1966, chap. 2. With permission.

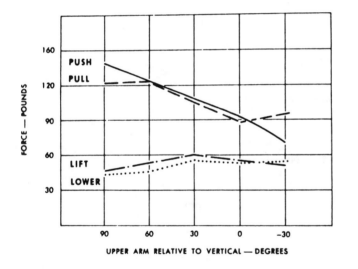
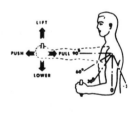

FIGURE 8.   Effects of pedal location on maximum force which can be applied. Up to 400 lb of force can be applied by an average man in area A of the graph, up to 600 pounds in area B. Line C represents a recommended optimum path of pedal travel where force application is considered a requirement.

Table 9
PEDAL PRESSURE ENDURANCE. TEST
CONDITIONS: KNEE ANGLE 111° ± 5°,
BACKREST ANGLE 13°, SEAT ANGLE 9°

| Pedal pressure exerted (lb) | Maximum endurance time — seconds | | | |
	Means	SD	5th	95th
			Percentiles	
200	218.5	122.8	17	421
300	114.6	62.9	11	218
400	68.4	31.7	16	121
500	39.6	21.2	5	75
600	20.1	12.8	—	41
700	11.7	7.8	—	25

From Damon, H., Stout, H. W., and McFarland, R., *The Human Body in Equipment Design,* Harvard University Press, Cambridge, 1966, chap. 2. With permission.

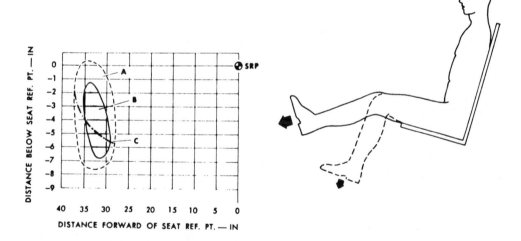

FIGURE 9.   Effect of upper arm angle on strength. (From Woodson, W. E. and Conover, D. W., *Human Engineering Guide for Equipment Design,* 2nd ed., University of California Press, Berkeley, 1964, chap. 5. With permission.)

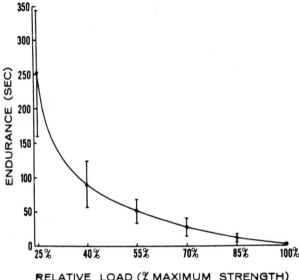

FIGURE 10.   Strength endurance. Duration of pull on an isometric dynamometer handle in front of the shoulder, at varying percentages of maximum exertable strength. Subjects were 36 college students, 18 males, 18 females. Abscissa indicates the percentage of each subject's maximum strength that he or she was asked to exert; ordinate indicates time in seconds that these forces could be maintained. Black dots shown mean values; vertical lines show ±1 SD. (From Damon, H., Stout, H. W., and McFarland, R., *The Human Body in Equipment Design,* Harvard University Press, Cambridge, 1966, chap. 2. With permission.)

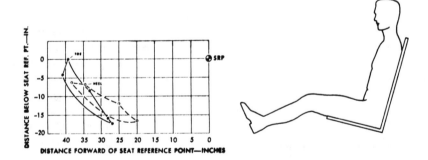

FIGURE 11.   Optimum areas for location of toe-operated and heel-operated pedal controls are shown in the above graph. (From Woodson, W. E. and Conover, D. W., *Human Engineering Guide for Equipment Design,* 2nd ed., University of California Press, Berkeley, 1964, chap. 5. With permission.)

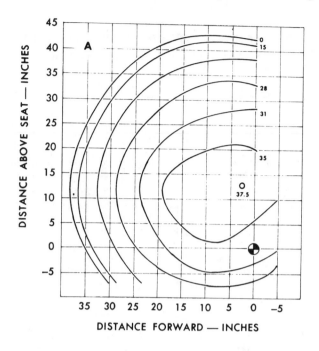

FIGURE 12. Arm reach for the 5th percentile unrestrained seated man in seven vertical planes located 0 to 37.5 in. from the subject's midline. (From Woodson, W. E. and Conover, D. W., *Human Engineering Guide for Equipment Design,* 2nd ed., University of California Press, Berkeley, 1964, chap. 5. With permission.)

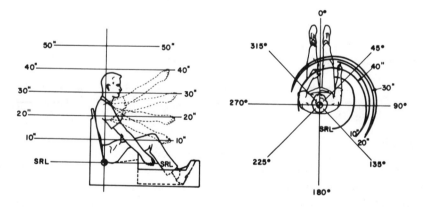

FIGURE 13. Grasping reach in several horizontal planes for an unrestrained seated 5th percentile subject. (From Damon, H., Stout, H. W., and McFarland, R., *The Human Party in Equipment Design,* Harvard University Press, Cambridge, 1966, chap. 2. With permission.)

Table 10
GRASPING REACH FOR THE RIGHT ARM, IN INCHES, TO A HORIZONTAL PLANE ABOVE THE SEAT REFERENCE POINT

| | Distance above seat, in. | | | | | | | | | |
| | 0 | | 10 | | 20 | | 30 | | 40 | |
Angle	5[a]	95[a]	5[a]	95[a]	5[a]	95[a]	5[a]	95[a]	5[a]	95[a]
0 Front					25.50	31.75	25.50	31.00	19.00	25.75
15					28.00	34.00	29.75	33.00	21.00	28.50
30	17.50	25.00	27.00	33.00	30.00	35.75	29.00	34.25	22.75	30.50
45	19.50	26.00	28.25	33.75	31.00	36.25	30.25	34.75	24.75	31.50
60	20.50	26.25	29.00	33.50	32.00	36.25	31.00	35.75	25.50	31.25
75	20.00	26.00	29.25	33.50	32.25	36.50	31.25	35.50	26.00	31.50
90	19.50	25.50	29.25	33.50	32.25	36.00	31.25	35.75	26.25	31.50
105	18.75	25.25	28.75	32.75	31.75	35.75	31.00	35.25	26.75	31.75
120	18.25	24.50	27.75	31.50	30.50	35.50	30.25	34.75	26.25	31.50
135	16.50	23.50	26.25	30.75		34.50		34.50		31.00
150	14.00	20.25		28.75						29.25
165		17.00						19.50		23.75
180								20.25		23.50
195								18.75		21.50
210								19.25		20.00
225								20.00		19.25
240								18.75	11.25	18.50
255								19.00	11.75	18.25
270				13.50		18.75		20.75	12.25	18.25
285				17.25		21.50		22.50	12.50	18.75
300				21.00	17.50	24.50	17.25	24.50	13.25	20.00
315				23.25	19.50	26.75	19.00	26.50	14.00	21.50
330				24.75	21.50	28.25	21.50	28.25	15.50	23.50
345				26.25	23.50	29.75	23.75	29.50	17.00	24.50

[a]   Percentiles.

From Kennedy, K. W., Reach capabilities of the UAAF population. Phase I, The Outer Boundaries of Grasping-Reach Envelope for Shirt Sleeved Operator, AMRL-TDR064-59, Aerospace Medical Research Laboratories, Wright-Patterson Air Force Base, Ohio, 1964. With permission.

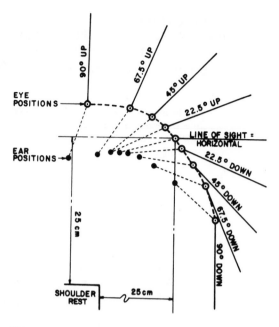

FIGURE 14.   Head position as defined by eye and ear location for straight ahead lines of sight from 90° down to 90° up. (From Damon, H., Stout, H. W., and McFarland, R., *The Human Party in Equipment Design,* Harvard University Press, Cambridge, 1966, chap. 2. With permission.)

Table 11
SELECTED DIMENSIONS OF THE HUMAN BODY (AGES 18 TO 45) SUITABLE FOR INITIAL DESIGN OF OPERATOR SPACE AND EQUIPMENT. LOCATIONS OF DIMENSIONS CORRESPOND TO FIGURE 15

Dimensional element	Men (in inches except where noted)		Women (in inches except where noted)	
	5th Percentile (132 lb)	95th Percentile (201 lb)	5th Percentile (102 lb)	95th Percentile (150 lb)
Standing				
Vertical reach	77.0-	89.0	69.0	81.0
Stature	65.0-	73.0	60.0	69.0
Eye to floor	61.0	69.0	56.0	64.0
Side arm reach from CL of body	29.0	34.0	27.0	38.0
Crotch to floor	30.0	36.0	24.0	30.0
Dimensions				
Chest circumference	35.0	43.0	30.0	37.0
Waist circumference	28.0	38.0	23.6	28.7
Hip circumference	34.0	42.0	33.0	40.0
Thigh circumference	20.0	25.0	19.0	24.0
Calf circumference	13.0	16.0	11.7	15.0
Ankle circumference	8.0	10.0	7.8	9.3
Foot length	9.8	11.3	8.7	10.2
Elbow to floor	41.0	46.0	34.0	46.0
Foot width	3.5	4.0	3.2	3.9
Head				
Head width	5.7	6.4	5.4	6.1
Interpupillary distance	2.27	2.74	1.91	2.94

Table 11 (continued)
SELECTED DIMENSIONS OF THE HUMAN BODY (AGES 18 TO 45)
SUITABLE FOR INITIAL DESIGN OF OPERATOR SPACE AND EQUIPMENT.
LOCATIONS OF DIMENSIONS CORRESPOND TO FIGURE 15

Dimensional element	Men (in inches except where noted)		Women (in inches except where noted)	
	5th Percentile (132 lb)	95th Percentile (201 lb)	5th Percentile (102 lb)	95th Percentile (150 lb)
Head length	7.3	8.2	6.4	7.3
Head height	—	10.2	—	9.0
Chin to eye	—	5.0	—	4.25
Head circumference	21.5	23.5	20.4	22.7
Hand				
Hand length	6.9	8.0	6.2	7.3
Hand width	3.7	4.4	3.2	4.0
Hand thickness	1.05	1.28	0.84	1.14
Fist circumference	10.7	12.4	9.1	10.7
Wrist circumference	6.3	7.5	5.5	6.9
Sitting				
Shoulder width	17.0	19.0	13.0	19.0
Sitting height to floor (std chair)	52.0	56.0	45.0	55.0
Eye to floor (std chair)	47.4	51.5	41.0	51.0
Standard chair	18.0	18.0	18.0	18.0
Hip breadth	13.0	15.0	12.5	15.4
Width between elbows	15.0	20.0	11.0	23.0
Workplace				
Arm reach (finger grasp)	30.0	35.0	22.0	33.0
Vertical reach	45.0	53.0	39.0	50.0
Head to seat	33.8	38.0	27.0	38.0
Eye to seat	29.4	33.5	25.0	32.0
Shoulder to seat	21.0	25.0	18.0	25.0
Elbow rest	7.0	11.0	4.0	12.0
Thigh clearance	4.8	6.5	3.5	6.0
Forearm length	13.6	16.2	14.0	18.0
Knee clearance to floor	20.0	23.0	17.0	22.0
Lower leg height	15.7	18.2	13.5	18.8
Seat length	14.8	21.5	13.0	23.0
Buttock-knee length	21.9	36.7	18.0	27.0
Buttock-toe clearance	32.0	37.0	27.0	37.0
Buttock-foot length	39.0	46.0	34.0	49.0

From Woodson, W. E. and Conover, D. W., *Human Engineering Guide for Equipment Design,* 2nd ed., University of California Press, Berkeley, 1964, chap. 5. With permission.

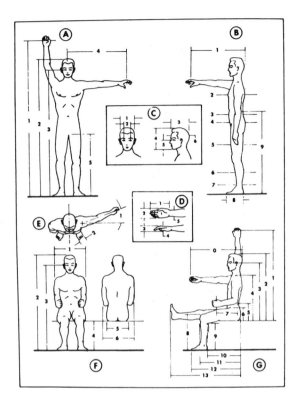

FIGURE 15. Outline of human body and location of the measurements reported in Table 11. (From Woodson, W. E. and Conover, D. W., *Human Engineering Guide for Equipment Design,* 2nd ed., University of California Press, Berkeley, 1964, chap. 5. With permission.)

REFERENCES

1. **Woodson, W. W. and Conover, D. W.,** *Human Engineering Guide for Equipment Designers,* 2nd ed., University of California Press, Berkeley, 1964, chap. 6.
2. **Fanger, P. O.,** *Thermal Comfort Analysis and Applications in Environmental Engineering,* Danish Technical Press, Copenhagen, 1970, chap. 2.
3. **Astrand, P. O. and Rodahl, K.,** *Textbook of Work Physiology,* McGraw-Hill, New York, 1970, chap. 15.
4. **Suggs, C. W.,** The role of enthalpy in heat loss, *Trans. Am. Soc. Agric. Eng.,* 9(3), 322, 1966.
5. **Suggs, C. W.,** Temperature and enthalpy equivalence of thermal radiation, *Trans. Am. Soc. Agric. Eng.,* 10(6), 727, 1967.
6. **Grandjean, E.,** *Fitting in the Task to the Man,* Taylor and Francis, London, 1969, 148.
7. **Lehmann, G.,** *Praktische Arbeitsphysiologie,* Thieme-Verlag, Stuttgart, 1953.
8. **Damon, H., Stout, H. W., and McFarland, R.,** *The Human Body in Equipment Design,* Harvard University Press, Cambridge, Mass., 1966, chap. 2.
9. **Woodson, W. E. and Conover, D. W.,** *Human Engineering Guide for Equipment Design,* 2nd ed., University of California Press, Berkeley, 1964, chap. 5.
10. **Kennedy, K. W.,** Reach capabilities of the UAAF population. Phase I, The Outer Boundaries of Grasping-Reach Envelope for Shirt-Sleeved Seated Operator, AMRL-TDRO64-59, Aerospace Medical Research Laboratories, Wright-Patterson Air Force Base, Ohio, 1964.

Plant Growth Data

PLANT GROWTH DATA

E. Dale Threadgill

Most plant growth data are very crop specific and exhibit tremendous variation with respect to soil type, climate, and season. Different cultivars of the same crop frequently have appreciably different growth characteristics.

The data in Table 1 provide depth of seeding, plant spacing, and seeding rate information for most crops.[1-3] The wide ranges of seed spacing reflect the dependence upon climate, season, intended market, etc. For agronomic crops, the lower seeding rates and wider spacings are more appropriate for nonirrigated production while higher rates and closer spacings apply for irrigated conditions. For vegetables, spacing is determined by type of market (processor vs. fresh market), desired size of fruit, harvesting equipment dimensions, season of year, etc. Depth of seeding for a given crop is determined by antecedent soil moisture, soil type, and soil temperature. The data in Table 1 can be utilized to establish general design criteria for planting and cultural equipment.

Evapotranspiration and the factors affecting it have been the subject of much research and discussion. Evapotranspiration (ET) has been found to depend upon the interactions between climatic factors, soil factors, fertility, and crop characteristics with respect to crop response. Crop water requirements depend upon ET with both factors having time-dependent distributions over the growth period of a crop.

Methods used to measure or estimate ET include lysimeter tanks, soil moisture depletion measurements, cut-plant weight loss, climatic data correlations, evaporation pan correlations, energy budget determinations, and mass transfer measurements. Excellent discussions and lists of references concerning these methods and the relationships between ET and crop characteristics, soil factors, and climatic factors have been developed by others.[4,5]

Estimates of ET for a particular crop obtained from these methods exhibit considerable variation. Errors between measured or estimated ET values and actual water requirements in field scale crop production cause great concern among scientists. There is sufficient disagreement about the validity or applicability of available seasonal ET values to preclude an acceptable tabulation of these data. Tabulations that are available in the literature should be used with adequate caution, since most data represent average values for either the entire U.S. or for a region. Pruitt and Doorenbos[6] discuss the need for empirical calibration of ET formulas and applying corrections to data for local conditions, especially when used in short-term prediction of crop water use such as required for irrigation management programs. Some data are reported as the ratio of ET for the crop of interest to the ET of a "standard" crop such as grass or alfalfa while other data are more general.[6-13] The cited references are indications of the type of information available and are not intended to be inclusive.

Table 1
SEED PLANTING DEPTH AND RATE

Crop	Depth to plant seed (cm)	Spacing between plants in row (cm)	Spacing between rows (cm)	Seeding rate (kg/ha)
Alfalfa	0.6—1.3	Drilled	Drilled	9—22
Asparagus	1.3	30—46	91—122	
Bahiagrass	1.3	Drilled	Drilled	11—13
Barley	1.3—2.5	Drilled	Drilled	81—108
Bean, snap bush	2.5—3.8	10—30	46—107	
Bean, Lima bush	2.5—3.8	10—20	46—107	
Bluegrass	1.3	Drilled	Drilled	17—28
Broccoli	0.6	36—61	61—91	
Cabbage	0.6	36—61	61—91	
Carrot	0.6	3—8	38—61	
Cauliflower	0.6	36—61	61—122	
Celery	0.6	15—30	46—91	
Collard	0.6	20—41	46—91	
Corn, Field	2.5—5.0	13—30	61—91	
Corn, Sweet	2.5—5.0	15—25	61—91	
Cotton	2.5—5.0	3—8	61—97	
Cucumber	2.5	30—91	91—122	
Kale	0.6	20—61	46—91	
Lespedeza	0.6—1.3	Drilled	Drilled	22—45
Lettuce, head	0.6—1.3	15—25	46—61	
Lettuce, leaf	0.6—1.3	5—8	38—46	
Lupine, blue	2.5—5.0	Drilled	Drilled	78—101
Millet	1.3—2.5	Drilled	Drilled	17—28
Muskmelon	2.5	30—91	152—244	
Mustard	1.3	15—30	30—61	
Oat	1.3—2.5	Drilled	Drilled	54—143
Okra	2.5	30—61	61—107	
Onion	1.3	8—10	38—91	
Parsley	0.6—1.3	15—20	38—91	
Pea, English	5.0	3—5	46—91	
Pea, southern	2.5	5—15	61—91	
Peanut	5.0—8.0	3—8	76—97	
Pepper	0.6	30—61	46—91	
Potato (tubers)	10.0	23—38	76—107	
Pumpkin	2.5—5.0	91—152	152—305	
Radish	1.3	3	30—61	
Rescuegrass	1.3—2.5	Drilled	Drilled	9—13
Rhubarb	10.0	61—122	91—152	
Rice	0.6—1.9	Drilled	Drilled	75—179
Rye	2.5	Drilled	Drilled	31—125
Ryegrass	1.3—2.5	Drilled	Drilled	28—34
Sorghum	1.3—5.0	Drilled	Drilled	17—50
Sorghum	1.3—5.0	5—13	60—91	
Soybean	2.5—3.8	3—15	30—91	
Spinach	2.0	5—15	30—91	
Squash, bush	2.5	46—122	91—152	
Sugarbeet	0.7—1.5	10—15	60—91	
Sunflower	2.5	41—61	91—122	
Sweet potato (plants)		30—41	91—122	
Tobacco (plants)		46—76	102—137	
Tomato, fresh	1.3	30—91	61—122	
Tomato, processor	1.3	23—61	61—122	
Turnip greens	1.3	5—8	30—61	

Table 1 (continued)
SEED PLANTING DEPTH AND RATE

Crop	Depth to plant seed (cm)	Spacing between plants in row (cm)	Spacing between rows (cm)	Seeding rate (kg/ha)
Vetch	2.5—3.8	Drilled	Drilled	22—90
Watermelon	2.5	61—244	183—244	
Wheat	1.5—2.5	Drilled	Drilled	67—101

REFERENCES

1. **Splittstoesser, W. E.,** Plant growth, in *Vegetable Growing Handbook,* AVI Publishing, Westport, Conn., 1979, 40.
2. **Knott, J. E.,** Plant growing and planting, in *Handbook for Vegetable Growers,* John Wiley & Sons, New York, 1957, 21.
3. **Martin, J. H., Leonard, W. H., and Stamp, D. L.,** in Principles of Field Crop Production, *3rd ed.,* Macmillan, New York, 1976, 1015.
4. **Evans, N. E., Taylor, S. A., Pruitt, W. O., Power, J. F., Evans, D. D., and Connell, G. H.,** *Water Requirements of Crops,* Spec. Publ. SP-SW-0162, American Society of Agricultural Engineers, St. Joseph, Mich., 1962, 56.
5. ASAE Conference Proceedings, Evaporation and its Role in Water Resources Management, *American Society of Agricultural Engineers,* Chicago, Ill., 1966, 65.
6. **Pruitt, W. O. and Doorenbos, J.,** Empirical Calibration, a Requisite for Evapotranspiration Formulae Based on Daily or Longer Mean Climatic Data?, in Proc. Int. Round Table Conf. on Evapotranspiration, International Commission on Irrigation and Drainage, Budapest, Hungary, 1977.
7. State of California, Vegetative Water Use in California, 1974, Bull. No. 113-3, Department of Water Resources, State of California, 1975, 104.
8. Rain Bird Sprinkler Mfg. Corp., Sprinkler Irrigation Handbook, Rain Bird Sprinkler Mfg. Corp., Glendora, Calif., 1971, 35.
9. **Thorne, D. W. and Thorne, M. D., Eds.,** Soil-plant-water relations, in *Soil, Water, and Crop Production,* AVI Publishing, Westport, Conn., 1979, 50.
10. **Thorne, D. W. and Thorne, M. D., Eds.,** Irrigation and crop production, in *Soil, Water, and Crop Production,* AVI Publishing, Westport, Conn., 1979, 97.
11. **Pruitt, W. O., Lourence, F. J., and Von Oettingen, S.,** Water use by crops as affected by climate and plant factors, *Calif. Agric.,* 26(10), 10, 1972.
12. Soil Conservation Service, Crop Consumptive Irrigation Coefficients and Irrigation Efficiency Coefficients for the United States, U.S. Department of Agriculture, Washington, D.C., 1976, 117.
13. **Stansell, J. R., Shepherd, J. L., Pallas, J. E., Bruce, R. R., Minton, N. A., Bell, D. K., and Morgan, L. W.,** Peanut responses to soil water variables in the Southeast, *Peanut Sci.,* 3(1), 44, 1976.

Irrigation Engineering

IRRIGATION WATER QUALITY

Glenn J. Hoffman

INTRODUCTION

Soluble salts are present in all natural waters, and it is their concentration and composition that determine the suitability of the water for irrigation. Irrigation water quality is normally based on three criteria: (1) salinity, the general effects of dissolved salts on crop growth that are associated with osmotic stress; (2) sodicity, the effect of an excessive proportion of sodium that induces a deterioration of soil structure; and (3) toxicity, the effects of specific solutes that damage plant tissue or cause an imbalance in plant nutrition. Irrigation water, however, has no inherent quality independent of the specific conditions under which it is to be used. Thus water quality can only be evaluated fully in the context of a specified set of conditions. This applies particularly to total salt content and chloride concentration. Other factors, such as sodium and boron concentrations, although influenced to some extent by crop tolerance and by soil characteristics, can often be evaluated without precise knowledge of the specific conditions. Thus only a generalized guide for evaluating irrigation water quality is presented because site-specific conditions may alter the suitability of a particular irrigation water.

COMPOSITION OF IRRIGATION WATER

The most prevalent cations in irrigation water are calcium, magnesium, sodium, and potassium; the typical anions are bicarbonate, sulfate, and chloride. Significant concentrations of other solutes, such as nitrate, carbonate, and trace elements are not common in natural waters. Nitrate, present in some localized areas, is beneficial to crop production unless it causes excessive vegetative growth or delays crop maturity. Nitrate, however, poses a potential health hazard to young children. Appreciable amounts of carbonate occur only infrequently when the pH of the water exceeds 8.5. Trace elements are not common in natural waters; but if present in only minute concentrations, each can severely limit production of certain crops. The concentration of trace elements may be a problem in waste waters used for irrigation.

Ion concentrations are typically reported in SI metric units as moles per cubic meter of solution (mol/m^3) or as grams per cubic meter (g/m^3). Traditionally, concentration has been expressed as milliequivalents per liter of solution (meq/ℓ) or as milligrams per liter (mg/ℓ). The units of gram per cubic meter (g/m^3) or milligrams per liter (mg/ℓ) are numerically equivalent to parts per million (ppm), an outmoded unit. To convert from mol/m^3 to g/m^3, multiply by the ion's atomic weight (given for convenience in Table 1). Calcium at a concentration of 10 mol/m^3 would be equivalent to a concentration of 401 g/m^3. The total salt concentration (C) is merely the sum of the concentration of each ion present.

The electrical conductivity, (σ) of an irrigation water is often measured to estimate C. For convenience, σ measurements are reported in SI metric units of decisiemens per meter (dS/m) or the traditional units of millimhos per centimeter (mmho/cm); the values are equal numerically. The relationship between salt concentration (g/m^3) and electrical conductivity can be approximated by $C = 640\ \sigma$. In addition to total salt content and the concentrations of individual constituents, the sodium-adsorption-ratio (R_{Na}) or an appropriately adjusted R_{Na} must be calculated to assess the sodium hazard of a water.[11] The determinations that are normally required to assess irrigation water quality are summarized in Table 1, along with their symbols and units of measure.

The quantities and types of salts present in irrigation water vary widely. Although the

Table 1
DETERMINATIONS NORMALLY REQUIRED TO
EVALUATE IRRIGATION WATER QUALITY ALONG
WITH THEIR SYMBOLS AND UNITS OF MEASURE

Determination	Symbol	Unit of measure	Atomic weight
Total salt content			
(1) Electrical conductivity	σ	dS/m[a]	—
(2) Concentration	C	mol/m^3 or g/m^3	—
Sodium hazard			
(1) Sodium adsorption ratio[b]	R_{Na}	(mol/m^3)$^{1/2}$	—
Constituents			
(1) Cations			
Calcium	Ca	mol/m^3	40.1
Magnesium	Mg	mol/m^3	24.3
Sodium	Na	mol/m^3	23.0
Potassium	K	mol/m^3	39.1
(2) Anions			
Bicarbonate	HCO$_3$	mol/m^3	61.0
Sulphate	SO$_4$	mol/m^3	96.1
Chloride	Cl	mol/m^3	35.5
Trace elements			
Boron	B	g/m^3	10.8

[a] dS/m = Decisiemens per meter = 1 millimho/cm, referenced to 25°C

[b] $R_{Na} = C_{Na}/\sqrt{C_{Ca} + C_{Mg}}$.

major sources of irrigation water are rivers and ground waters, use of brackish waters and waste waters will increase as supplemental irrigation and land disposal of municipal and industrial waste waters increase. The average salt content of rivers, world-wide, is estimated at about 120 g/m^3,[6] but the range is large. Rivers used for irrigation generally contain more salt than the average because they occur in arid regions, although some in the Pacific Northwest and California remain low in salinity for much of their lengths. For comparison, compositions of representative river waters are given in Table 2. Ground waters are generally more saline than surface waters and commonly contain higher proportions of sodium, boron, and nitrate. Changes in water quality with pumping duration are not common, but significant differences among relatively closely spaced wells are frequently encountered. Such changes reflect the composition of different strata from which water is being pumped. For example, wells within 1 km of each other in the Coachella Valley of California have been reported to have salt concentrations varying from 0.6 to 12 dS/m. Typical water analyses of irrigation wells are also given in Table 2.

SALINITY ASSESSMENT

Most irrigation waters do not contain enough salt to cause immediate injury to crops, although one exception is the foliar damage caused by sprinkling sensitive crops with saline water under high evaporative conditions. Following an irrigation, however, the concentration of the soluble salts in the soil increases as water is removed by evaporation and transpiration, leaving the salt behind. Without leaching, salt accumulates in the soil with successive irrigations. With leaching, the accumulation of soluble salts in the soil is controlled by the fraction of the applied water that drains below the crop rootzone (leaching fraction). Hence, the salinity assessment must be made in view of the salt tolerance of the crop and the leaching fraction. The fraction of salt in the irrigation water that will precipitate is a further consideration if the leaching fraction is less than about 0.1.[12]

Table 2
WATER QUALITY OF SELECTED RIVER AND WELL WATERS

Water source and location	Salinity		Sodicity	Constituents							
	Electrical conductivity (dS/m)	Salt content (g/m³)	R_{Na} (mol/m³)$^{1/2}$	Cations				Anions			B (g/m³)
				Ca (mol/m³)	Mg (mol/m³)	Na (mol/m³)	K (mol/m³)	HCO₃ (mol/m³)	SO₄ (mol/m³)	Cl (mol/m³)	
River waters[a]											
San Joaquin Friant Dam, Calif.	0.04	20	0.4	0.07	0.02	0.12	0.02	0.2	0.02	0.05	—
Snake King Hill, Idaho	0.5	370	0.9	1.1	0.7	1.2	0.1	3.3	0.5	0.7	—
Rio Grande Falcon Dam, Tex.	1.0	680	2.9	1.8	0.8	4.6	0.1	2.2	2.4	3.2	—
Colorado Hoover Dam, Ariz.	1.3	920	3.2	2.3	1.4	6.2	0.1	3.0	3.5	3.6	0.2
Gila Gillespie Dam, Ariz.	4.4	2920	10.7	4.8	3.5	30.7	0.3	4.5	6.7	28.4	2.1
Pecos Artesia, N.M.	8.6	6010	12.5	12.1	8.5	56.8	0.3	2.7	17.7	60.2	0.6
Well waters[b]											
Indio, Calif.	0.3	230	1.4	0.7	0.2	1.3	0.0	2.2	0.3	0.2	0.02
Bakersfield, Calif.	0.8	530	23.1	0.1	0.0	7.3	0.0	2.5	1.2	2.5	6.9
Scottsdale, Ariz.	1.2	820	3.9	1.6	1.4	6.8	—	3.2	1.6	6.2	—
Tolleson, Ariz.	4.0	2870	14.6	3.2	2.0	33.3	—	10.8	4.0	24.9	—
Pecos, Tex.	4.4	2860	6.1	4.6	8.0	21.8	0.7	1.9	8.4	29.1	—
Roll, Ariz.	7.2	4540	13.2	7.4	5.8	48.0	0.3	6.3	6.6	55.4	1.3

[a] Data are averages for the 1978 or 1979 water year from U.S. Geological Survey (1979, 1980).
[b] Data taken from Hoffman et al.[4]

Agricultural crops vary widely in their salt tolerance (see Table 3); thus there is a range of permissible salinity levels in irrigation water, other factors being equal. In addition, varietal and rootstock variations occur among some crop species, and tolerance may also vary with plant growth stage.[7] At soil salinity levels higher than the threshold value for a given crop (see Table 3), crops generally show a progressive decrease in size with increasing salinity. Fruit and seed yield does not always parallel vegetative response to salinity. Thus crop characteristics and crop management can influence the level of permissible salinity in the irrigation water. Climate has also been shown to modify plant response to salinity. Generally, salinity effects are more pronounced under hot, dry conditions than under cool, humid ones; but not all species are affected equally.[7]

A generalized appraisal of the salinity hazard of an irrigation water can be made based upon proper water management and the ability of the soil to meet or exceed the minimum leaching fraction required to maintain full crop production, termed the leaching requirement. The leaching requirement as a function of irrigation water quality and crop salt tolerance is given in Figure 1. This relationship allows the prediction of the minimum leaching fraction required for the water and crop in question that avoids any loss in yield. For example, if the salinity of the applied water, amount-weighted average of the salinity of the irrigation water and rainfall, was 2 dS/m and the crop was alfalfa that has a salt tolerance threshold of 2 dS/m (Table 3), the leaching requirement would be 0.18. If the soil or the proposed water management would not provide for at least 18% of the applied water to leach through the root zone, alfalfa yield would be reduced by salination. Furthermore, without at least 18% leaching, this quality of water could not be used indefinitely without converting to a more salt tolerant crop or reclaiming the soil profile periodically.

SODICITY ASSESSMENT

Another consideration in evaluating irrigation water quality is the potential for an excess concentration of sodium in the soil to deteriorate soil structure. When calcium and magnesium are the predominant cations adsorbed on the soil exchange complex, the soil tends to have a granular structure that is easily tilled and readily permeable. When the amount of adsorbed sodium exceeds about 10% of the total cation-exchange capacity of the soil, however, soil mineral particles tend to disperse and water penetration decreases. Excess sodium becomes a problem when the rate of infiltration is reduced to the point that the crop is not adequately supplied with water or when the hydraulic conductivity of the soil profile is too low to provide adequate drainage. Sodium may also add to cropping difficulties through crusting of seed beds, temporary saturation of surface soil, and/or possible disease, weed, oxygen, nutritional, and salinity problems.

The sodium-adsorption-ratio (R_{Na}) of the irrigation water is generally a good indicator of the exchangeable sodium status that will occur in the soil. R_{Na} is defined as

$$R_{Na} = C_{Na}/\sqrt{(C_{Ca} + C_{Mg})}$$

where all ion concentrations (C) are in mol/m^3. In some cases, R_{Na} of the irrigation water must be adjusted to account for the influence of carbonates and bicarbonates on the precipitation of calcium and magnesium.[11]

The value of R_{Na} when a permeability problem may occur, is influenced by the type of clay minerals present in the soil.[1] In general, however, no permeability problem should be expected if the R_{Na} is below 10 (mol/m^3)$^{1/2}$, providing the irrigation water is not too pure. For the well water from Bakersfield, California (see Table 2), the R_{Na} is 23.1 and permeability problems should be anticipated.

Table 3
SALT TOLERANCE OF AGRICULTURAL CROPS
AS A FUNCTION OF THE ELECTRICAL
CONDUCTIVITY OF THE SOIL SATURATION
EXTRACT (σ_e) WHERE RELATIVE YIELD (Y) IN
PERCENT = 100 − B(σ_e − A)^7 and $\sigma_e \geqslant$ A

Crop	Salt tolerance threshold[a] (A)	Percent yield decline[b] (B)	Qualitative salt tolerance rating[c]
	(dS/m)[d]	%/(dS/m)	
Alfalfa	2.0	7.3	MS
Almond	1.5	19	S
Apple	—	—	S
Apricot	1.6	24	S
Avocado	—	—	S
Barley (forage)	6.0	7.1	MT
Barley (grain)	8.0	5	T
Bean	1.0	19	S
Beet, garden	4.0	9	MT
Bentgrass	—	—	MS
Bermudagrass	6.9	6.4	T
Blackberry	1.5	22	S
Boysenberry	1.5	22	S
Broadbean	1.6	9.6	MS
Broccoli	2.8	9.2	MS
Bromegrass	—	—	MT
Cabbage	1.8	9.7	MS
Canarygrass, reed	—	—	MT
Carrot	1.0	14	S
Clover	1.5	12	MS
Clover, berseem	1.5	5.7	MS
Corn (forage)	1.8	7.4	MS
Corn (grain)	1.7	12	MS
Corn (sweet)	1.7	12	MS
Cotton	7.7	5.2	T
Cowpea	1.3	14	MS
Cucumber	2.5	13	MS
Date palm	4.0	3.6	T
Fescue, tall	3.9	5.3	MT
Flax	1.7	12	MS
Grape	1.5	9.6	MS
Grapefruit	1.8	16	S
Hardinggrass	4.6	7.6	MT
Lemon	—	—	S
Lettuce	1.3	13	MS
Lovegrass	2.0	8.4	MS
Meadow foxtail	1.5	9.6	MS
Millet, foxtail	—	—	MS
Okra	—	—	S
Olive	—	—	MT
Onion	1.2	16	S
Orange	1.7	16	S
Orchardgrass	1.5	6.2	MS
Peach	1.7	21	S
Peanut	3.2	29	MS
Pepper	1.5	14	MS
Plum	1.5	18	S

Table 3 (continued)
SALT TOLERANCE OF AGRICULTURAL CROPS
AS A FUNCTION OF THE ELECTRICAL
CONDUCTIVITY OF THE SOIL SATURATION
EXTRACT (σ_e) WHERE RELATIVE YIELD (Y) IN
PERCENT = $100 - B(\sigma_e - A)^7$ and $\sigma_e \geq A$

Crop	Salt tolerance threshold[a] (A)	Percent yield decline[b] (B)	Qualitative salt tolerance rating[c]
Potato	1.7	12	MS
Radish	1.2	13	MS
Raspberry	—	—	S
Rhodesgrass	—	—	MS
Rice, paddy	3.0	12	MS
Ryegrass, perennial	5.6	7.6	MT
Safflower	—	—	MT
Sesbania	2.3	7	MS
Sorghum	—	—	MS
Soybean	5.0	20	MT
Spinach	2.0	7.6	MS
Strawberry	1.0	33	S
Sudangrass	2.8	4.3	MT
Sugarbeet	7.0	5.9	T
Sugar cane	1.7	5.9	MS
Sweet potato	1.5	11	MS
Timothy	—	—	MS
Tomato	2.5	9.9	MS
Trefoil, big	2.3	19	MS
Trefoil, birdsfoot	5.0	10	MT
Vetch	3.0	11	MS
Wheat	6.0	7.1	MT
Wheatgrass, crested	3.5	4	MT
Wheatgrass, fairway	7.5	6.9	T
Wheatgrass, slender	—	—	MT
Wheatgrass, tall	7.5	4.2	T
Wildrye, altai	—	—	T
Wildrye, beardless	2.7	6	MT
Wildrye, Russian	—	—	T

[a] Salt tolerance threshold is the mean soil salinity at initial yield decline.
[b] Percent yield decline is the rate of yield reduction per unit increase in salinity beyond the threshold.
[c] Qualitative salt tolerance ratings are sensitive (S), moderately sensitive (MS), moderately tolerant (MT), and tolerant (T).
[d] Salinity is expressed as the electrical conductivity of a soil saturation extract σ_e (dS/m = decisiemens per meter = 1 millimho per centimeter referenced to 25°C).

From Maas, E. V. and Hoffman, G. J., *J. Irrig. Drainage Am. Soc. Civil Eng.*, 103 (IRZ), 115, 1977. With permission.

Water of very low salt content can aggravate a permeability problem because it allows a maximal swelling and dispersion of soil minerals and organic matter and because it has a tremendous capacity to dissolve and remove calcium. The addition of gypsum to the dilute irrigation water in parts of the San Joaquin Valley of California is a common practice to improve infiltration. This problem can arise if the electrical conductivity of the irrigation

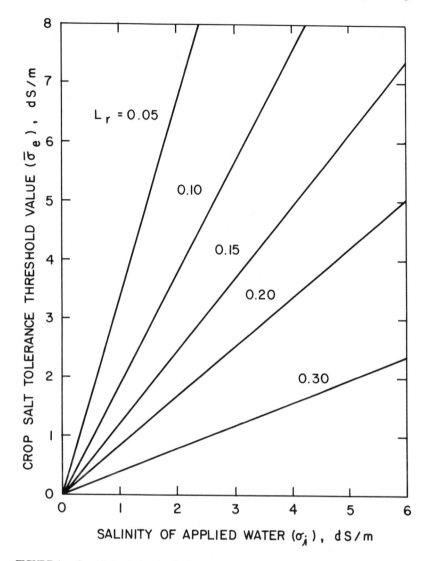

FIGURE 1. Graphical solution for the leaching requirement as a function of the salinity of the applied water and the salt tolerance threshold value for the crop. (From Hoffman, G. J., and van Genuchten, M. Th., Soil properties and efficient water use: water management for salinity control, 1983. With permission.)

water is less than 0.2 dS/m. Irrigating with water from the San Joaquin River (see Table 2) may cause problems on soils with potential permeability problems.

TOXICITY ASSESSMENT

The most common constituents of irrigation water that may be toxic to crops are chloride, sodium, and boron. The reader is referred elsewhere[8] for an in-depth discussion of trace elements like selenium and metals that may be toxic.

Trees and other woody perennials, unlike most annual crops, may be specifically sensitive to chloride and sodium. Crop, varietal, and rootstock differences in tolerance to these ions depends largely upon the rate the ion is transported from the soil to the leaves. Leaf injury symptoms appear in chloride-sensitive crops when leaves accumulate 0.25 to 0.70% chloride

Table 4
CHLORIDE TOLERANCE OF FRUIT CROP ROOTSTOCKS
AND VARIETIES IF LEAF INJURY IS TO BE AVOIDED

		Maximum permissible chloride concentration	
		---	---
Crop	Rootstock or variety	Soil saturation extract[c]	Plant leaf analysis[b]
	Rootstocks	(mol/m³)	(kg/kg)
Citrus	Rangpur lime		
Citrus spp.	mandarin	25	0.7
	Rough lemon, tangelo		
	sour orange	15	
	Sweet orange, citrange	10	
Stone fruit	Marianna	25	0.3
Prunus spp.	Lovell, Shahl	10	
	Yunnan	7	
Avocado	West Indian		
Persea americana	Mexican	8	0.25—0.50
Grape	Salt creek, 1613-3	40	0.5
Vitis spp.	Dog ridge	30	
	Varieties		
Grape	Thompson seedless, Perlette	25	0.5
Vitis Spp.	Cardinal, Black Rose	10	
Olive			0.5
Olea europaea			
Berries			
Rubus spp.	Ollalie blackberry	10	
	Indian summer raspberry	5	
Strawberry	Lassen	8	
Fragaria spp.	Shasta	5	

[a]   From Bernstein.[2]
[b]   From Reisenauer.[10]

on a dry-mass basis. Symptoms develop as leaf burn or drying of leaf tissue, typically occurring first at the extreme tip of older leaves and progressing back along the leaf edges. Excessive leaf burn is often accompanied by defoliation. Chemical analysis of soil or leaves can be used to confirm chloride toxicity. The maximum permissible concentrations of chloride in the soil saturation extract or in plant leaves for several sensitive crops are given in Table 4.

Symptoms of sodium toxicity occur first on older leaves as a burning or drying of tissue at the outer edges of the leaf. As severity increases, the affected zone progresses toward the center of the leaf between the veins. For many tree crops, sodium concentration in leaf tissue of 0.2 to 0.5% (dry-mass basis) are the maximum allowable without injury symptoms. Sodium toxicity is often modified and reduced if calcium is present. Because of this interaction, a reasonable evaluation of the potential toxicity is given by the exchangeable-sodium-percentage (ESP) of the soil.[15] ESP is the proportion of the soil exchange complex occupied by sodium. The tolerance of representative crops to sodium is given as a function of soil ESP in Table 5.

Irrigation methods that wet plant leaves, such as overhead sprinkling, may cause specific

Table 5
TOLERANCE OF VARIOUS CROPS TO EXCHANGEABLE-SODIUM-
PERCENTATE (ESP) UNDER NONSALINE CONDITIONS[9]

Tolerance to ESP and range at which affected	Crop	Growth response under field conditions
Extremely sensitive (ESP = 2—10)	Deciduous fruits Nuts Citrus *Citrus* spp. Avocado *Persea americana*	Sodium toxicity symptoms even at low ESP values
Sensitive (ESP = 10—20)	Bean *Phaseolus vulgaris*	Stunted growth at these ESP values even though soil physical conditions may be good
Moderately tolerant (ESP = 20—40)	Clover *Trifolium* spp. Oat *Avena sativa* Tall Fescue *Festuca elatior* Rice *Oryza sativa* Dallisgrass *Paspalum dilatatum*	Stunted growth due to nutritional factors and adverse soil conditions
Tolerant (ESP = 40—60)	Wheat *Triticum aestivum* Cotton *Gossypium hirsutum* Alfalfa *Medicago sativa* Barley *Hordeum vulgare* Tomato *Lycopersicon* Beet *Beta vulgaris*	Stunted growth usually due to adverse soil physical conditions
Most tolerant (ESP = more than 60)	Crested and Fairway wheatgrass *Agropyron* spp. Tall wheatgrass *Agropyron elongatum* Rhodesgrass *Chloris gayana*	Stunted growth usually due to adverse soil physical conditions

From Pearson, G. A., Tolerance of Crops to Exchangeable Sodium, USDA Inf. Bull. 216, U.S. Department of Agriculture, Washington, D.C., 1960.

ion toxicity problems at sodium or chloride concentrations lower than those that cause problems with surface irrigation methods. This occurs primarily during periods of high temperature and low humidity. Excess chloride and sodium can accumulate in leaves by foliar absorption, and the more frequent the wetting and drying cycles, the greater the leaf damage. For example, citrus, sprinkler-irrigated in several California valleys, has been damaged with chloride and sodium concentrations as low as 3 mol/m^3. These same concentrations caused no toxic effects with furrow or flood irrigation.[3]

Boron, although an essential minor element, is phytotoxic if present in excess. Most boron toxicity problems arise from high concentrations in well waters or springs located near geothermal areas or geological faults. Few surface waters contain sufficient boron to cause toxicity. Sensitivity to boron is not limited to woody perennials, but affects a wide variety of crops. Boron toxicity symptoms typically appear at the tip and along the edges of older leaves as yellowing, spotting, and/or drying of leaf tissue. The damage gradually progresses interveinally toward midleaf. A gummosis or exudate on limbs or trunks is sometimes noticeable on boron-affected trees, such as almond. Many sensitive crops show toxicity symptoms when boron concentrations in leaf blades exceed 250 mg/kg, but not all sensitive crops accumulate boron in their leaves. Stonefruits (e.g., peach, plum, almond) and pome fruits (pear, apple, and others) may not accumulate boron in leaf tissue to the extent that leaf analysis is a reliable toxocity indicator.

A wide range of crops have been tested for boron tolerance in sand cultures. The results of these tests have been summarized in Table 6 by grouping the crops according to their

Table 6
RELATIVE TOLERANCE OF CROPS TO BORON

Tolerant[a] 4.0 g/m³ of boron	Semitolerant 2.0 g/m³ of boron	Sensitive 1.0 g/m³ of boron
Asparagus *Asparagus officinalis*	Sunflower, native *Helianthus annuus*	Pecan *Carya illinoinensis*
Date palm *Phoenix dactylfera*	Potato *Solanum tuberosum*	Walnut, black and Persian, or English *Juglans* spp.
Sugarbeet *Beta vulgaris*	Cotton, acala and pima *Gossypium spp.*	Jerusalem artichoke *Helianthus tuberosus*
Garden beet *Beta vulgaris*	Tomato *Lycopersicon esculentum*	Navy bean *Phaseolus vulgaris*
Alfalfa *Medicago sativa*	Radish *Raphanus sativus*	Plum *Prunus domestica*
Broadbean *Vicia faba*	Field pea *Pisum sativum*	Pear *Pyrus communis*
Onion *Allium cepa*	Olive *Olea europaea*	Apple *Malus sylvestris*
Turnip *Brassica oleracea*	Barley *Hordeum vulgare*	Grape *Vitis* spp.
Cabbage *Brassica oleracea* Capitata	Wheat *Triticum aestivum*	Kadota fig *Ficus carica*
Lettuce *Lactuca sativa*	Corn *Zea mays*	Persimmon *Diospyros virginiana*
Carrot *Daucus carota*	Sorghum *Sorghum bicolor*	Cherry *Prunus* spp.
	Oat *Avena sativa*	Peach *Prunus persica*
	Pumpkin *Cucurbita* spp.	Apricot *Prunus armeniaca*
	Bell pepper *Capsicum annuum*	Thornless blackberry *Rubus* spp.
	Sweetpotato *Ipomoea batatas*	Orange *Citrus sinensis*
	Lima bean *Phaseolus lunatus*	Avocado *Persea americana*
		Grapefruit *Citrus x. paradisi*
		Lemon *Citrus limon*
2.0 g/m³ of boron	1.0 g/m³ of boron	0.3 g/m³ of boron

Note: Relative tolerance is based on the boron concentration in irrigation water at which boron toxicity symptoms were observed when plants were grown in sand culture. It does not necessarily indicate a reduction in crop yield.

[a] Tolerance decreases in descending order in each column between the stated limits.

Adapted from Wilcox, L.V., Boron Injury to Plants, USDA Bull. 211, U.S. Department of Agriculture, Washington, D.C., 1960.

relative boron tolerance. These data are based on the boron level at which toxicity symptoms were observed, and do not necessarily indicate reductions in yield.

SUMMARY

The suitability of a given water for irrigation is dependent upon the crop, soil, environment, and water management. Once these conditions are known, the various water quality criteria can be evaluated. If soil and water management conditions permit the required leaching fraction, a salinity problem is unlikely. If the required leaching is not met, a more salt-tolerant crop must be grown or management must change to avoid salination. The threat of creating sodic soil conditions can be determined from the ion concentrations of the water and soil mineralogy. To avoid possible toxic effects from various solutes in the water, maximum acceptable levels for a given crop should not be exceeded. The guidelines given here have been generalized and some moderation may be required for site-specific conditions.

REFERENCES

1. **Ayers, R. S. and Wescott, D. W.,** Water Quality for Agriculture. FAO Irrig. and Drain. Paper 29, Food and Agriculture Organization, Rome, 1976.
2. **Bernstein, L.,** Salt Tolerance of Fruit Crops, USDA Inf. Bull. 292, U.S. Department of Agriculture, Washington, D.C., 1965.
3. **Harding, R. B., Miller, M. P., and Fireman, M.,** Absorption of salts by citrus leaves during sprinkling with water suitable for surface irrigation, *Proc. Am. Soc. Hortic. Sci.,* 71, 248, 1958.
4. **Hoffman, G. J., Ayers, R. S., Doering, E. J., and McNeal, B. L.,** Salinity in irrigated agriculture, in *Design and Operation of On-Farm Irrigation Systems,* Jensen, M. E., Ed., American Society of Agricultural Engineers Monograph, St. Joseph, Mich., 1980, 145.
5. **Hoffman, G. J. and van Genuchten, M. Th.,** Soil properties and efficient water use: water management for salinity control, in *Limitations to Efficient Water Use in Crop Production,* Taylor, H. M., Jordan, W. R., and Sinclair, T. R., Eds. American Society of Agronomy, Madison, Wisc., 1983, 73.
6. **Livingstone, D. A.,** Chemical composition of rivers and lakes, in Data of Geochemistry, 6th ed., Geol. Survey Prof. Paper 440-6, Fleischer, M., Ed., U.S. Geological Survey, Washington, D.C., 1963.
7. **Maas, E. V. and Hoffman, G. J.,** Crop salt tolerance — current assessment, *J. Irrig. Drainage Am. Soc. Civil Eng.,* 103(IRZ), 115, 1977.
8. National Academy of Sciences and National Academy of Engineering, Water Quality Criteria, EPA-R3-73-033, U.S. Government Printing Office, Washington, D.C., 1972.
9. **Pearson, G. A.,** Tolerance of Crops to Exchangeable Sodium, USDA Inf. Bull. 216, U.S. Department of Agriculture, Washington, D.C., 1960.
10. **Reisenauer, H. M.,** Soil and Plant — Tissue Testing in California, Univ. of Calif. Bull. 1879, University of California, 1976.
11. **Rhoades, J. D.,** Quality of water for irrigation, *Soil Sci.,* 113, 277, 1972.
12. **Rhoades, J. D., Oster, J. D., Ingvalson, R. D., Tucker, J. M., and Clark, M.,** Minimizing the salt burdens of irrigation drainage waters, *J. Environ. Quality,* 3, 311, 1974.
13. U.S. Geological Survey, Water Resources Data, Water Year 1978, Geological Survey Water-Data Rep. CA-78-1, CA-78-3, AZ-78-1, U.S. Geological Survey, Washington, D.C., 1979.
14. U.S. Geological Survey, Water Resources Data, Water Year 1979. Geological Survey Water-Data Rep. ID-79-1, TX-79-3, NM-79-1, U.S. Geological Survey, Washington, D.C., 1980.
15. U.S. Salinity Laboratory Staff, Diagnosis and improvement of saline and alkali soils, USDA Agriculture Handbook 60, U.S. Department of Agriculture, Washington, D.C., 1954.
16. **Wilcox, L. V.,** Boron Injury to Plants, USDA Bull. 211, U.S. Department of Agriculture, Washington, D.C., 1960.

IRRIGATION WELL DESIGN AND SPECIFICATIONS

James H. Poehlman
and
Jay H. Lehr

INTRODUCTION

This chapter will aid in writing specifications for the design and construction of an adequate agricultural irrigation supply well. While industrial and municipal water supply wells typically are designed by consulting engineers, most irrigation wells have been designed by local water well contractors after a discussion with the irrigator. While this has been satisfactory for the farmer who needs to irrigate only a few acres of crops or needs frost protection, the demands put on a well by a large commercial grower can exceed the capacities of these ''shade-tree engineering'' projects. The use of the specifications found in this chapter will assist both the grower and the water well contractor. By inserting specific parameters into the appropriate specification blanks, the final design and construction methods will provide guidelines for the well contractor to construct the most efficient well possible under the existing conditions.

The design and construction of any supply well depends upon the geologic conditions and the eventual use of the well. The requirements placed upon a water supply well vary greatly. The most diverse set of demands are those of an irrigation well. When a commercial or municipal supply well is constructed, the demands are relatively constant. That is, the well is designed to yield a specific amount of water at a continuous flow. This particular design is usually specified by a consultant or an engineer. However, an irrigation well is used intermittently and at varying flow rates. This irregular usage dramatically increases maintenance demands. Constant flow and regular use will, in some cases, actually assist in the development of a well; the irregular flow pattern of an irrigation well can often disturb the formation and in some instances cause aquifer failure.

The time a well is unused is also detrimental to the performance of the well. When a well is idle for several months, the silt and dissolved solids that naturally occur in ground water have an opportunity to settle. This settling can cause plugging of the formation and screen.

Variable use distinguishes an irrigation well from its commercial and municipal counterparts. Aside from this, the design and construction are nearly identical.

Before an irrigation well is drilled, there are several questions to be answered. Although the order is not critical, the following items must be determined.

1. Type of crop to be irrigated — to establish quality limitations as well as quantity demands
2. Area to be irrigated — to determine quantity demand
3. Type of irrigation to be used — to determine quality limitations as well as quantity
4. Flow requirements — to determine construction dimensions
5. Availability of required flow-gathered from test wells or from local geological records — to determine if the requirements are available from existing aquifers; this will also contribute to dimensions and number of wells required

Once these factors are known, the design criteria to be applied to the well or wells can be established. These will include:

1.   Anticipated capacity of the well in gallons per minute (gal/min) or liters per second ($\ell \ sec^{-1}$).
2.   Number of wells to be constructed
3.   Casing diameter
4.   Diameter and length of screen
5.   Construction method
6.   Development method
7.   Method and duration of testing
8.   Payment method
9.   Warranties

Upon completion of all preliminary studies and based upon minimum design criteria, the construction contract can be written. The owner or engineer should not tell the contractor how to construct the supply well, but rather should establish construction guidelines and precise requirements for the material to be used in the construction of the well.

The specifications should be divided into two separate parts: general conditions and special conditions.

General conditions will include the legal agreement between the owner and the contractor and establish such items as work schedules, subcontract standards, substitution allowances, change orders, and payment schedules. It will also define the rights and requirements of both the owner and the contractor.

Special conditions will define the agreed criteria for the completion of the well, recognizing that some variation may occur. This section will define each specific project.

TEST WELL

Description
The contractor shall furnish all labor, materials, tools, equipment, power, and water and all other services necessary for drilling, casing, and development of a test well. The test well will be located in an accessible area and the boring shall be used to determine the design and depth of the supply well.

Depth
The test well shall be drilled through the water-bearing formation or formations. It is anticipated that the depth will be ____ feet (meter). However, this depth may vary according to geologic conditions at the drill site.

Drilling
The test well shall be constructed plumb and straight in an approved manner using a water well rig of sufficient strength and of standard manufacture. The contractor shall have a competent operator in charge and present at the site during the drilling operations.

Logs and Samples
Accurate driller's logs of all materials penetrated shall be recorded by a driller who must be experienced and have certifiable skills to determine the depths and thickness of the strata encountered.

Driller's Log
During the drilling of the test hole the contractor shall prepare and keep a complete log consisting of the following:

1. The reference point for all depth measurement (feet or meters above sea level)
2. The depth of each change of formation
3. The identification of the material of each formation
4. The depth at which water was first encountered
5. The depth from which water and formation sample is taken
6. The diameter of the hole
7. The depth to Static Water Level (SWL)
8. Total depth of well
9. Depth and dimension of casing
10. Description and location of screen or type, size and location of perforations
11. Any and all other pertinent information for a complete and accurate log, e.g., temperature, pH, and results of water sample tests

Samples
Formation Samples
Formation samples are to be taken starting at ____ feet (meters) and at each 5-ft (2 m) interval or formation change thereafter. Special care is to be exercised in collection of samples from suspected producing zones.

Water Samples
Water samples shall consist of sufficient quantity to provide complete and accurate testing by an approved water testing laboratory. The chemical analysis shall include the following determinations.[1]

1. Boron
2. Calcium
3. Magnesium
4. Sodium
5. Bicarbonate
6. Sulfate
7. Nitrate
8. Total dissolved solids
9. Potassium
10. Carbon dioxide
11. Sodium adsorption rates
12. ____,____,____

Disposition
All test wells are to be either converted to observation wells or abandoned.

Conversion
Conversion to observation wells consists of installing a minimum 2 in. (5-cm) standard weight[2] steel pipe or S.D.R. 21[3] PVC pipe and a minimum 3 ft (1 m) of screen or perforations. Each observation well shall terminate a minimum of 2 ft (0.6 cm) above grade and be closed with a water-tight threaded cap or suitable equivalent.

Abandonment
Abandonment shall conform to Article 56.000-000-000 of the Manual of Water Well Construction Practices by the U.S. Environmental Protection Agency, Office of Water Supply.[1]

WATER SUPPLY WELL

Description

The contractor shall furnish all labor, materials, tools, equipment, power, water, and all other services necessary for drilling, casing, and developing a water supply well. The well shall be located in an accessible area as determined by the test well program.

Depth

The supply well shall be drilled through the water bearing formation as defined by the test well program. It is anticipated that the well will be approximately ____ feet (meters) deep. However, this depth may vary according to geologic conditions encountered at the drill site.

Drilling

The supply well shall be constructed plumb and straight in an approved manner using a water well rig of sufficient strength, in good operating condition, and of standard manu-facture. The contractor shall have a competent operator in charge and present at the site during the drilling operations.

Permits

After the award of the contract, the contractor shall acquire all required permits from authorized agencies. No field operation shall commence until these approvals have been obtained.

Logs and Samples

Accurate driller's logs of all materials penetrated shall be recorded by a driller who must be experienced and have certifiable skills to determine the depths, thickness, and charac-teristics of the strata encountered.

Driller's Log

During the drilling of the test hole, the contractor shall prepare and keep a complete log consisting of the following:

1. The reference point for all depth measurement (feet or meters above sea level)
2. The depth of each change of formation
3. The identification of the material of each formation
4. The depth at which water was first encountered
5. The depth from which water and formation sample is taken
6. The diameter of the hole
7. The depth to static water level (SWL)
8. The total depth of well
9. Depth and dimension of casing
10. Description and location of screen or type, size and location of perforations
11. Any and all other pertinent information for the complete and accurate log, e.g., temperature, pH, and appearance of water samples

Sample
Formation Samples

Formation samples are to be taken starting at ____ feet (meters) and at each 5-ft (2 m) interval or formation change thereafter. Special care is to be exercised in collection of samples from suspected producing zones.

Water Samples

Water samples shall consist of sufficient quantity to provide complete and accurate testing by an approved water testing laboratory. The chemical analysis shall include the following determinations:[1]

1.   Boron
2.   Calcium
3.   Magnesium
4.   Sodium
5.   Potassium
6.   Carbon dioxide
7.   Bicarbonate
8.   Sulfate
9.   Nitrate
10. Total dissolved solids
11. Sodium adsorption rates
12. ____, ____, ____,

Geophysical Logging

Geophysical logging would include all techniques involving the lowering of sensing devices into the borehole and recording physical parameters to interpret the characteristics of the rocks, the fluids contained in the rocks, or the construction of the well. It shall be the contractor's responsibility to conduct any geophysical logging operation called for by the specifications or to arrange for it unless otherwise directed by the owner or his agent.

The following geophysical/mechanical logs are to be conducted:____, ____, ____, ____, _____. These logs shall meet all appropriate specifications as defined in Article 45, Preamble, of the Manual of Water Well Construction Practices by the U.S. Environmental Protection Agency, Office of Water Supply.[1]

Materials

The materials used in the construction of the supply well shall be in accordance with the following requirements.

Drilling Fluids

Materials used by the contractor to prepare the drilling fluid shall be composed of fresh, nonpolluted water and sodium bentonite type drilling clay that has been commercially processed to meet or surpass the viscosity specifications of the American Petroleum Institute Standard 13-A for Drilling Fluid Materials.[4] All other drilling fluid additives used will comply with recognized industry standards and practices, and they will be applied and used as prescribed by the manufacturer. All parties understand that toxic and/or dangerous substances will not be added to the drilling fluid.

The contractor shall be responsible for maintaining the quality of the drilling fluid to assure: (1) protection of the water bearing formations and potential water bearing formations exposed in the borehole, and (2) good representative samples of the formation materials.

The contractor is responsible for the complete removal of the drilling fluid from the hole and the development of the well, as per the well development specifications.

Casing

All well casing shall be new and must conform to appropriate standards such as American Petroleum Institute (API), American Society for Testing and Materials (ASTM), and American Water Works Association (AWWA). All casing shall bear mill markings that will identify the material as that which is specified. All joints should be made in a manner suitable to the material to ensure that they will be watertight where necessary. If steel casing is welded, the standards of the American Welding Society should apply.

1.   Steel well casing shall meet the standards of the AWWA Specification Al-5.4, Table 1 Weights of Steel and Wrought Iron Casing Pipe[5] or API Standard 13A2 and the latest revisions thereto.
2.   Polyvinylchloride (PVC). Plastic well casing shall be new pipe conforming with ASTM F-480-76 and latest revisions thereto. Joining of PVC casing must also conform to the assembly procedures established and set forth in ASTM F-480-76.[3]

Grouting

Grouting consists of filling the annulus or other space with an impervious material. The reasons for grouting are

1.   Protection of the aquifer, or aquifers, including the prevention of water movement between aquifers, for purposes of maintaining quality or preserving the hydraulic response of the producing zone(s)
2.   Protecting the well against the entry of water from the surface or a subsurface zone
3.   Protection of casing; this may be necessary to guard against attack by corrosive waters, or where special assurance of structural integrity is desired

A satisfactory grouting program must result in complete envelopment of the casing, in keeping with the following guidelines.

Neat Cement Grout

A mixture of Portland cement (ASTM C-150), and not more than 7 gal of clean water per bag (1 ft^3 or 94 lb) of cement, shall be used. The use of special cements, or bentonite to reduce shrinkage or other admixtures (ASTM C494) to reduce permeability, increase fluidity, and/or control time to set, and the composition of the resultant slurry must be approved by the owner or his agent.

Concrete Grout (Alternate)

A mixture of Portland cement (ASTM C-150) sand, coarse aggregate, and water in the proportion of at least five bags of cement per cubic yard of concrete to not more than 7 gal of clean water per bag of cement (1 ft^3 or 94 lb) shall be used. The use of cement, or bentonite to reduce shrinkage or other admixtures (ASTM C494) to reduce permeability, increase fluidity, and/or control time to set, and the composition of the resultant slurry must be approved by the owner or his agent.

Installation

Grouting shall be continuous from the bottom of the permanent casing to the land surface; or, where a filter pack has been installed, from the top of the pack (following development) to the land surface; or where a well screen only has been installed, from a point 5 ft (2 m) above the screen to the land surface. The actual method of placement shall be selected from Article 48.000-000-000, of the Manual of Water Well Construction Practices.[1] No further work shall be done on the well until the grout has firmly set (a minimum of 72 hr for neat cement or concrete and 24 hr for quick setting cement).

Well Screen

The well screen aperture openings, screen length, and diameter should be selected so as to have sufficient open area to transmit the desired yield with an aperture (slot) entrance velocity equal to or less than 6 ft (2 m)/min (0.1 ft (0.03 m)/sec).

Louvered Pipe

The screen shall consist of a pipe that has punched openings in it where the material has not been removed. The openings formed shall be between the corner of the outside of the pipe and the punched-out area, and the corner of the inside of the punched portion and its side.

Continuous Slot Wire Wound Screen

The screen shall be constructed of wound wire, reinforced with longitudinal bars, the bars having a cross section that will form an opening between each adjacent coil of wire that is shaped in such a manner as to increase in size inward. The wire shall be firmly attached to the bars that will, in turn, be attached to a coupling adapter.

Material and Fittings

To reduce the possibility of corrosion, the well screen and its end fittings should be fabricated of the same material. The choice should be made on the basis of chemical analysis of the water or prior knowledge of the water quality. Some possible choices include:

1. Type 304 stainless steel
2. Silicon bronze
3. Everdur metal (96% copper, 3% silicon, 1% manganese)
4. Silicon red brass
5. Armco iron
6. Mild steel
7. PVC
8. Reinforced fiberglass

The screen(s) shall be set at an elevation, or elevations in multiple zones, that approximate the best producing zone or zones. The selection of such settings should be based on results of an analysis of the formation penetrated as recorded in the driller's log, stratigraphic log, and any geophysical/mechanical logs available.

Setting and Screen

The screen, with a closed bottom, shall be attached to the casing by an approved manner and lowered into the well with the casing. In no instance shall it be driven or forced. It shall be suspended from the surface until the formation has collapsed against it or until a filter pack has been installed.

Filter Material (Gravel Pack)

The filter pack should consist of clean, well-rounded grains that are smooth and uniform. The filter should be siliceous with a limit of 5% by weight of calcareous materials. The grain size and gradation of the filter are selected to stabilize the aquifer material and to permit only the fine factions to move into the well during development. Thus, after development, a correctly filtered well is relatively sand free, and a narrow annulus of the formation adjacent to the filter has its permeability increased to some degree.

Filter Grain Size

The filter grain size and gradation shall be the contractor's responsibility. The thickness of the filter pack shall range from a minimum of 3 in. (8 cm) to approximately 8 in. (20 cm).

Length of Filter

Filter media shall be placed in the annulus between the casing or screen and side of the borehole from the bottom of the well to a point approximately 5 ft (2 m) above the top of the uppermost screen. The filter shall be placed by the tremie pipe method. The tremie pipe shall be controlled so that the bottom of the pipe is not more than 5 ft above the filter media in the well. The filter shall be installed with a quantity of water sufficient to flush the filter from the pipe to the bottom of the well to prevent segregation.

Plumbness and Alignment

The casing and screen shall be set plumb regardless of any out-of-plumb condition of the borehole. The completed well shall be sufficiently plumb and straight so that there will be no interference with the installation, alignment, operation, or future removal of the permanent well pump. The contractor, at his own expense, shall make the test for plumbness at the discretion of the owner. If a test for plumbness is required, it shall be performed according to Section 51.200-000-000 of the Manual of Water Well Construction Practices.[1]

Development

The supply well shall be fully developed by extracting all of the sand possible by acceptable agitation methods. It is anticipated that various procedures will be required to fully develop the wells, and it is not intended for these specifications to give a full description of all procedures that may be required. The contractor shall perform his work in accordance with accepted development practices such as those described in Article 52.000-000-000 of the Manual of Water Well Construction Practices.[1]

Sand Content

Well development shall continue until ____ppm sand content is achieved as measured according to Article 52.000-001-000 or until ordered to stop by the owner or his agent.

Payment for Development

All work performed under this section shall be paid for at the unit price bid per hour for well construction equipment plus the cost of any materials and/or subcontracted services at the invoice price plus a percent of add-on bid:[1] (1) well development _____ unit price/hr, and (2) chemicals _____ unit price/lb (kg).

Performance Testing

Data obtained from pumping will provide information necessary to determine well capacity, aquifer characteristics, well efficiency, pumping rates, pump installation depth settings, and other factors that will be of value in the long-term operation and maintenance of the well. Design of the well should be based upon facts derived from careful analysis of data from properly conducted tests. Comprehensive aquifer tests require a minimum of two observation wells, depending on the purpose of the results from the well tests.

Tests

For conducting pumping tests, the contractor shall furnish test pumping equipment capable of pumping 200% of the anticipated flow rate. The contractor shall furnish the driver for the pump, discharge piping, throttling valves, flow measurement device, water level recording apparatus, and any other equipment required to perform the pumping tests. The pumping test shall be of the step-continuous composite method. The contractor shall furnish, install, and remove the necessary measuring instruments and pumping equipment capable of pumping to the required point of discharge a minimum of ____ gal/min (ℓ/sec) with the pumping level ____ feet (m), and with satisfactory throttling devices, so that the discharge may be reduced to ____ gal/min (ℓ/sec). The pumping unit shall be complete with an ample power source, controls, and appurtenances and shall be capable of being operated without interruption for a period of ____ hours.

Prior to starting the pump, water level measurements shall be made at least hourly, for a minimum of 8 hr, in the production well and all observation wells, and these measurements shall be recorded on the same note sheets to be used during the pumping test. The well shall be "step" tested at rates of approximately one half, three fourths, one, and one and a half times the design capacity of ____ gal/min (ℓ/sec). The complete test is estimated to require approximately ____ hours. The contractor shall operate the pump and change the discharge as directed by the owner. Discharge of the pump shall be controlled by a gate valve, if electric driven, and both gate valve and engine throttle if engine driven. The discharge shall be controlled and maintained at approximately the desired discharge for each step with an accuracy of plus or minus 5%. Pump discharge shall be measured with a totalizing meter and stopwatch, circular orifice meter, or Venturi meter as approved by the owner. A 1/4-in. (15 cm) tube free of air leaks shall be installed in the well with the test pump, terminating 5 ft (1.5 m) above the pump intake. The tube shall have an accurate altitude gauge and air valve attached to it at the surface. The vertical distance from the bottom of the air line to the center of the gauge shall be recorded. The line shall then be pressurized to at least 1 lb/in^2 (0.07 atm) for each 2 ft of air (0.6m) line and until the gauge will read no higher. The water level in the well will be computed by subtracting the altitude registered in feet on the gauge from the air line's length. This method is not recommended for small drawdowns due to its lack of precision.[1]

Well Disinfection

The contractor shall provide for disinfection as soon as construction of the well, performance testing, and cleaning procedures have been completed. The contractor shall carry out adequate cleaning procedures immediately preceding disinfection where evidence indicates that normal well construction and development work have not adequately cleaned the well. All oil, grease, soil, and other materials, which could harbor and protect bacteria from disinfectants, shall be removed from the well. Unless prior approval is obtained for employing chemicals or unusual cleaning methods, the cleaning operation is to be carried out by pumping and swabbing only. Where test pumping equipment is to be utilized, such equipment shall be installed prior to or during disinfection and be thoroughly hosed, scrubbed, or otherwise cleaned of foreign material.[1]

Water Sample

Water samples shall consist of sufficient quantity to provide complete and accurate testing by an approved water testing laboratory. The chemical analysis shall include the following determinations:[1]

1. Boron
2. Calcium
3. Magnesium
4. Sodium
5. Potassium
6. Carbon dioxide
7. Bicarbonate
8. Sulfate
9. Nitrate
10. Total dissolved solids
11. Sodium absorption rates
12. ____, ____, ____

Well Abandonment

Aquifer sealing criteria: aquifers shall be filled with disinfected, dimensionally stable materials and compacted mechanically if necessary to avoid later settlement. (Cement, cement-and-sand, and concrete do not require disinfection.)

Disinfection of aquifer fill materials shall be accomplished by using chlorine compounds such as sodium hypochlorite or calcium hypochlorite. Aquifer fill materials shall be clean (relatively free of clays and organic materials) before placement in the well. Disinfection shall be accomplished by dissolving sufficient chlorine compound to produce a calculated concentration of at least 100 ppm available chlorine in double the volume of water in the well. The fill material shall be placed in the well after the water in the well has been so treated. Nonproducing zones above the aquifer shall be filled with stable materials such as sand, sand-and-gravel, cement, cement-and-sand, or concrete. Nonproducing zones above the uppermost aquifer seal shall be filled with materials less permeable than the surrounding undisturbed formations. The upper most 5 ft (2 m) of the bore hole (to land surface) shall be filled with a material appropriate to the intended use of the land.[1]

Table 1
BID SCHEDULE, TEST WELL

Item no.	Work or material	Quantity and unit	Unit price	Amount
1.	Mobilization and demobilization	For the lump sum of		$_____
2.	Drilling_____ in. diam. hole	_____ft	_____lin ft	
3.	Geophysical/mechanical log(s)	For the lump sum of		$_____
4.	Furnishing and installing_____ test casing including_____ ft of well screen	_____ft	_____ft	$_____
5.	Developing test wells	_____hr	$_____hr	$_____hr
6.	Furnishing, installing test pump	For the lump sum of		$_____
7.	Lump sum price using estimated quantities			$_____

Table 2
BID SCHEDULE, GRAVEL PACKED IRRIGATION WATER WELL

Item no.	Work or material	Quantity and unit	Unit price	Amount
1.	Mobilization and demobilization	For the lump sum of		$_____
2.	Drilling _____in. diam. hole to accommodate outer casing	_____lin ft	$_____/ft	$_____
3.	Furnishing and installing _____in. diam. outer casing	_____lin ft	$_____/ft	$_____
4.	Grouting_____ diam. outer casing	_____Sacks of cement	$_____/ sack	$_____
5.	Drilling _____diam. hole beneath outer casing for screen and inner casing	_____lin ft	$_____/ft	$_____
6.	Geophysical/mechanical logs	For the lump sum of		$_____
7.	Furnishing and installing _____in. diam. inner casing with centering guides	_____lin ft	$_____/ft	$_____
8.	Furnishing and installing _____in. diam. well screen with fittings and centering guides	_____lin ft	$_____/ft	$_____
9.	Furnishing and installing gravel pack	_____cu ft	$_____/ft	$_____
10.	Developing the well	_____hr	$_____/hr	$_____
11.	Furnishing and installing seal between inner and outer casing	For the lump sum of		$_____
12.	Furnishing, installing, and removing test pump	For the lump sum of		$_____
13.	Testing the well via pumping with test pump	_____hr	$_____/hr	$_____
14.	Disinfecting and capping well	For the lump sum of		$_____
15.	Standby time	_____hr	$_____/hr	$_____
16.	Lump sum price using estimated quantities			$_____

GLOSSARY

Abandoned Well — A well whose original purpose and use have been permanently discontinued or which is in such a state of disrepair that its original purpose cannot be reasonably achieved.

Agreement — The written agreement between the owner and contractor, as indicated by the Bid and Bonds.

Air Line — The smaller vertical air pipe usually submerged to within a few feet of the bottom of the eductor pipe. The length of the air line below the static water level is used in calculating the air pressure required to start the air-lift.

Annular Space — The space between the casing and the well bore or casing surrounding it.

Aquifer — An aquifer is a formation, group of formations, or part of a formation that contains sufficient saturated permeable material to yield significant quantities of water to wells and springs.

Area of Influence — The area beneath which groundwater, or pressure-surface contours, are modified by pumping.

Artesian — Artesian is synonymous with confined. Artesian water and artesian water body are equivalent, respectively, to confined ground water and confined water body. An artesian well is a well deriving its water from an artesian or confined water body. The water level in an artesian well stands above the top of the artesian water body it taps.

Artesian Well — A well tapping a confined or artesian aquifer in which the static water level stands above the water table. The term is sometimes used to include all wells tapping confined water, in which case those wells with water level above the water table are said to have positive artesian head (pressure) and those with water level below the water table, negative artesian head.

Bacteria — The nuisance organisms developing and multiplying in wells, which do not cause disease, commonly known as *Clonothrix, Crenothrix, Gallionella, Leptothrix, Siderocapsa,* and *Sphaerotilus* of the aerobic type; Sulfur Bacteria and Sulfate-Reducing Bacteria of the anaerobic type. Also Iron Bacteria.

Bentonite — A highly plastic colloidal clay largely of mineral mortmorillonite.

Calcium — The most frequent cause of hardness and affects the scale-forming and corrosive properties of water.

Casing — A tubular retaining structure, generally metal, which is installed in the evacuated hole to maintain the well opening.

Casing Shoe — A heavy-walled steel coupling or band at the lower extremity of the casing.

Cement — The powdered-dry cement prior to the addition of the mixing water.

Cementing — The process of placing the cement slurry to provide a seal against subsurface waters.

Centralizer — Used to center casing in the hole, insuring uniform annular space for effective grouting.

Chloride — The chlorides of calcium, magnesium, sodium, iron, etc. normally found in water are extremely soluble. In natural waters, high chloride may indicate animal pollution, but usually is from passing through a salt formation in the earth of sea water intrusion. A salty or brackish taste may be imparted to the water.

Clay — A soft, plastic variously colored earth composed largely of hydrous silicate of alumina, formed by the decomposition of feldspar and other aluminum silicates.

Contractor — The person, firm, or corporation with whom the owner has executed the Agreement.

Drag Bit — Equipped with short blades and a body fitted with water courses that direct the drilling fluid stream to keep the blades clean. Assists penetration by means of a jetting action against the bottom of the hole.

Drawdown — Lowering of water level caused by pumping. It is measured for a given quantity of water pumped during a specified period, or after the pumping level has become constant.

Dry Hole or Duster — A well drilled which produced neither oil nor gas, nor water of significant quantity.

Fish — Any foreign material in a well that cannot be removed at will.

Fishing — The act of attempting to recover a fish.

Formation — An assemblage of rock masses grouped together into a unit that is convenient for description or mapping.

Fracture — A break. Fracture is a general term to include any kind of discontinuity in a body of rock if produced by mechanical failure whether by shear-stress or tensile stress. Fractures include faults, shears, joints, and planes of fracture-cleavage.

Geophysical/Mechanical Logging — Geophysical logging is composed of a number of techniques that measure some electrical, chemical, or radioactive property of the subsurface, either characteristic of the ground water or of the rocks in which the ground water occurs. Typical techniques include: Resistivity and Self-potential Logging (called Electric Logging), Gamma and Neutron Logging (called Radiation Logging), etc. Mechanical logging incorporates mechanical devices, as opposed to electronic or electric, that measure some physical property of the subsurface, e.g., Caliper Logging, Temperature Logging, Photographic Logging, etc.

Gravel Packed Well — A well in which filter material is placed in the annular space to increase the effective diameter of the well, and to prevent fine-grained sediments from entering the well.

Ground Water — Water in the zone of saturation.

Head (static) — The height above a standard datum of the surface of a column of water (or other liquid) that can be supported by the static pressure at a given point.

Head (total) — The sum of three components: (1) elevation head, h_e, which is equal to the elevation of the point above a datum, (2) pressure head, h_p, which is the height of a column of static water that can be supported by the static pressure at the point, and (3) velocity head, h_v, which is the height the kinetic energy of the liquid is capable of lifting the liquid.

Hydrologic Cycle — All movements of water and water vapor in the atmosphere, on the ground surface, below the surface, and return to the atmosphere by evaporation and transpiration.

Hardness — Calcium and magnesium salts, present in the most natural waters, cause the water to be hard. Natural water will vary in hardness depending upon its location. When the total hardness exceeds the carbonate and bicarbonate alkalinity, the amount of hardness equivalent to the alkalinity is carbonate hardness and the remainder is noncarbonate hardness. When the carbonate and bicarbonate alkalinity equal or exceed the total hardness, all hardness is carbonate.

Hydrology — A science concerned with the occurrence of water in the earth, its physical and chemical reactions with the rest of the earth, and its relation to the life of the earth.

Iron — In the ferric state, iron is completely oxidized, but in the ferrous state it is only partially oxidized. It exists in most natural waters in the ferrous state. Exposure to air or addition of chlorine oxidizes the iron to the ferric state and it may hydrolyze to form insoluble hydrated ferric oxide. The form of iron may also be changed by iron bacteria. Hence the iron may be in true solution; in colloidal state; complexed by phosphates; as relatively coarse suspended particles; and as ferrous, ferric, or both. Release of carbon dioxide will convert iron from the ferrous to the ferric state. It may be placed into solution when corrosion of iron and steel surfaces occurs.

Laminar Flow — Motion of a fluid the particles of which move substantially in parallel paths. This type of flow always occurs below the lower critical velocity and may occur between the lower and higher critical velocities. Also called straight-line, stream-line, or viscous flow.

Loss of Circulation — The loss of drilling fluid into formation pores or crevices.

Nitrogen (Nitrate) — The most highly oxidized phase in the nitrogen cycle. In surface water supplies, occurrence is generally in trace quantities, but some ground waters may have higher quantities.

Nitrogen (Nitrite) — An intermediate stage in oxidation or reduction process of the nitrogen cycle in water. Trace amounts in raw surface waters indicates pollution. It may also be produced by action of bacteria or other organisms on ammonia nitrogen supplied at elevated temperatures in combined residual chlorination of water.

Nitrogen (Organic) — Amino acids, polypeptides, proteins and albuminoid nitrogen contribute to the organic nitrogen content of the water. A rise in the organic nitrogen content may indicate sewage or industrial waste pollution.

Organic Polymers — Used in drilling fluids and have a strong affinity for water, (i.e., they are hydrophilic colloids). They develop highly swollen gels, are absorbed by clay particles, and protect the clay from the flocculating effects of salts.

Oxygen (Dissolved) — Present in all surface waters and rain waters due to contact with the atmosphere, it is found in lesser quantities in well waters and may be absent in deep well supplies. Its presence creates a corrosive effect on iron and steel.

Packer — A device placed in a well that plugs or seals the well at a specific point.

Perched Ground Water — Ground water in a saturated zone that is separated from the main body of ground water by unsaturated rock.

Perforations — A series of openings in a well casing, made either before or after installation of the casing, to permit the entrance of water into the well.

pH — A measure of the acidity or alkalinity of water.

Phosphate — In many natural waters, phosphate occurs in trace amounts. Raw or treated sewage, agricultural drainage, and some industrial waters contain significant concentration of phosphate. Both ortho and polyphosphates may be found in the same sample. Trace amounts may be combined with organic matter. Phosphate analysis is made primarily to control chemical dosage or to trace flow or contamination.

Porosity — The property of a rock or soil containing interstices or voids and may provide for fluid transmission. It is expressed as a percentage of the total volume occupied by the interconnecting interstices. Although effective percentage with respect to the movement of water only the system of interconnected interstices is significant.

Potentiometric Surface — Replaces the term "piezometric surface", is a surface which represents the static head. As related to an aquifer, it is defined by the levels to which water will rise in tightly cased wells. Where the head varies appreciably with depth in the aquifer, a potentiometric surface is meaningful only if it describes the static head along a particular specified surface or stratum in that aquifer. More than one potentiometric surface is then required to describe the distribution of head. The water table is a particular potentiometric surface.

Sand Content — The percentage bulk volume of sand in a drilling fluid.

Saturation Zone — The zone below the water table in which all interstices are filled with ground water.

Sodium — Present in most natural waters and in fairly high concentrations in water softened by the sodium exchange process. A high sodium to cations ratio can be detrimental to soil permeability in agriculture.

Soil Moisture — Pellicular water of the soil zone. It is divided by the soil scientist into available and unavailable moisture. Available moisture is water easily abstracted by root action and is limited by field capacity and the wilting coefficient. Unavailable moisture is water held so firmly by adhesion or other forces that it cannot usually be absorbed by plants rapidly enough to produce growth. It is commonly limited by the wilting coefficient.

Specific Capacity — The rate of discharge of water from the well divided by the drawdown of water level within the well. It varies slowly with duration of discharge, which should be stated when known. If the specific capacity is constant except for the time variation, it is roughly proportional to the transmissivity of the aquifer. The relation between discharge and drawdown is affected by the construction of the well, its development, the character of the screen or casing perforation, and the velocity and length of flow up the casing. If the well losses are significant, the ratio between discharge and drawdown decreases with increasing discharge; it is generally possible to roughly separate the effects of the aquifer from those of the well by step drawdown tests. In aquifers with large tubular openings, the ratio between discharge and drawdown may also decrease with increasing discharge because of a departure from laminar flow near the well, or in other words, a departure from Darcy's law.

Specific Yield — The ratio of (1) the volume of water which the rock or soil, after being saturated, will yield by gravity to (2) the volume of the rock of soil. The definition implies that gravity drainage is complete. In the natural environment, specific yield is generally observed as the change that occurs in the amount of water in storage per unit area of unconfined aquifer as the result of a unit change in head. Such a change in storage is produced by the draining or filling of pore space and is therefore dependent upon particle size, rate of change of the water table, time, and other variables. Hence, specific yield is only an approximate measure of the relation between storage and head in unconfined aquifers. It is equal to porosity minus specific retention.

Standing Level — The water level in a nonpumping well. The term is used without regard to whether the well is within or outside the area of influence of pumping wells. If outside the area of influence, the term is equivalent to static level; if within the area of influence, the standing level registers one point on the cone of pumping depression.

Submergence — The length of air pipe submerged below the pumping level divided by the sum of the submerged length and the lift, and quotient being multiplied by 100 to give the result as percent. Generally, a submergence of 60% or more is desirable.

Sulfide — Most commonly, sulfide is found in well waters and is a result of bacterial action on organic matter under anaerobic conditions. Hydrogen sulfide imparts a "rotten egg" odor to the water, is toxic, and imparts a corrosive character to the water.

Test Well — A well completed for pumping and/or sampling.

Tremie Pipe — A device, usually a small diameter pipe, that carries grouting materials to the bottom of the hole and which allows pressure grouting from the bottom up without introduction of appreciable air pockets.

Turbidity — Suspended matter such as clay, silt, finely divided organic matter, plankton, and other microscopic organisms cause turbidity in water. The standard method for determination is the Jackson candle turbidimeter.

Water Table — The surface in an unconfined water body at which the pressure is atmospheric. It is defined by the levels at which water stands in wells that penetrate the water body just far enough to hold standing water. In wells which penetrate to greater depths, the water level will stand above or below the water table if an upward or downward component or ground water flow exists.

Well Efficiency — The actual specific capacity adjusted for well loss, divided by the theoretical specific capacity.

Well Log — See Geophysical/Mechanical Logging.

Well Screen — Serves as the intake section of the well that obtains water from an aquifer of unconsolidated materials such as sand. It allows water to flow freely into the well from water saturated sand, prevents sand from entering with the water, and serves as a structural retainer to support the bore hole in unconsolidated material. Numerous types are available and their application depends on the specific hydrogeologic conditions present.

REFERENCES

1. EPA, Manual of Water Well Construction Practices, 570/9-75-001, National Water Well Association Research Department, U.S. Environmental Protection Agency, Washington, D.C., 1975.
2. API, *API Specifications for Casing, Tubing and Drill Pipe,* A.P.I. Standard 5A, 31st ed., American Petroleum Institute, 1971.
3. Annual Book of Standards, Part 34, *Standard Specification for Thermoplastic Water Well Casing Pipe and Coupling Made in Standard Dimension Ratios (SDR),* Subcommittee F17.61, American Society for Testing and Materials, Philadelphia, 1977.
4. API, *A.P.I. Standard 13-A for Drilling Fluid Materials,* A.P.I. Committee on Standardization of Drilling Fluid Materials, American Petroleum Institute, 1969.
5. AWWA, *AWWA Standard for Deep Wells A1-058,* Water Works Practice Committee 631OR, American Water Works Association, New York, 1958.
6. NWWA, *Proposes Specification for Large Capacity Rotary Drilled Gravel Packed Wells,* NWWA Specifications Committee, National Water Well Association, Worthington, Ohio.

SPRINKLER SYSTEM DESIGN

Claude H. Pair

INTRODUCTION

Sprinkler systems are composed of lateral pipe lines and main pipe lines that conduct water under pressure from a pumping plant or gravity source to the sprinklers. The sprinklers apply water in the form of spray like rainfall to the soil or other surface. The spray is developed by discharging water under pressure through sprinkler nozzles or perforated pipe. The water pressure may be developed by gravity or pumping. The systems are used for agricultural and turf irrigation, land disposal of liquid waste, environmental control, application of fertilizer, weed control chemicals, or insecticides, and for fire protection. The largest use is in irrigation for agriculture and turf.

Designers of sprinkler systems for irrigation use must combine information on the size and shape of the area to be irrigated, the topography, the cropping pattern if more than one crop will be irrigated, the amount of water used by each crop, the soils, climate, water supplies, power supplies, labor, and management schedules to produce the most economical and efficient system.

The first step in the system design is to make an inventory of the available resources, operating conditions, and desires of the user. A scale topographic map showing field boundaries of areas to be irrigated, roads, buildings, and other obstructions, soil types and depths, water supply location quantity and quality, power supply type and location gives much of the basic design information. The crops and crop rotation for the irrigated area, the labor supply and management problems all affect a system type and its operation and have to be considered in its design. Then there are the wishes of the purchaser of the system as to whether a completely portable system, a semiportable system, or permanent system is wanted.

The completely portable system may have a fixed or portable pumping plant with portable main lines and portable lateral pipe lines from the water source to the last sprinkler. A portable pipe line is one that can easily be manually connected or disconnected and reassembled at other locations. This type system is the most economical from the capital investment standpoint, but requires the most operating labor.

The semiportable systems have both permanently located and portable pipe lines. The water source is usually permanent with permanently located main pipe line from it to the lateral pipe lines, which are usually portable. The permanent main line pipe may be buried underground.

The permanent sprinkler systems have fixed location water sources and permanently located main line pipes with permanently located lateral piping to conduct water to each sprinkler on the irrigated area. This is the most expensive irrigation system and requires the least operating labor.

DETERMINATION OF NET AND GROSS WATER APPLICATION

With all the information for a particular area gathered, the design of the system proceeds by determining the net and gross depth of water to apply at each irrigation. This is a function of the soil water-holding capacity, the soil moisture level requirement of the crop, and the water application efficiency of the sprinkler system.

The soil water-holding capacity is a function of the type and root zone depth for the crop being irrigated, discussed in chapters by Talbot and Treadgills. The application efficiency of the sprinkler system is the ratio of the depth of water stored in the soil to depth applied by the sprinkler system. The water loss in a well-designed and operated system occurs because of wind drift and evaporation between sprinkler and ground. The uneven water distribution is due to sprinkler nozzle pressure changes that occur because of irregular irrigated topography and pipe friction loss. The efficiency used in design varies from 60% in a hot, windy, dry air area to 80% in a cool, less wind, and very humid air area.

The net amount of water to be stored per irrigation is calculated using the equation:

$$d = S \times W_h \qquad (1)$$

where d is the depth of water to be stored in millimeters (inches), S is root zone soil depth in meters (feet), and W_h is the water-holding capacity in millimeters per meter (inches per foot).

The gross amount of water to be applied each irrigation is calculated by dividing the net depth stored (d) by the irrigation efficiency. See chapter by Wu et al. The formula is:

$$D = \frac{d}{E} \qquad (2)$$

where D is the gross depth of water applied in millimeters (inches), d is the net depth of water applied in millimeters (inches), and E is the irrigation efficiency in percent.

TIME BETWEEN IRRIGATIONS

The maximum time allowed between irrigations is the depth of water stored per irrigation divided by the crop maximum peak daily use rate. This is obtained by the following formula:

$$F = \frac{d}{P_u} \qquad (3)$$

where F is the irrigation interval in days, d is the depth of water stored in the root zone of crop in millimeters (inches), and P_u is the peak water use rate of the crop in millimeters (inches) per day. This maximum time period between irrigations should be determined for each crop and for each season.

SYSTEM CAPACITY REQUIREMENTS

The next step is to determine the system capacity requirements at crop peak use rates. This depends on the size of the irrigated area, the gross depth of water applied at each irrigation, and the operating time allowed to apply this water depth. The system capacity is computed by the formula:

$$Q = K \frac{AD}{FH} \qquad (4)$$

where Q is the sprinkler system capacity in liters (gallons) per minute, K is a constant depending on the measurement units, 166.8 metric or 453 for U.S., A is the design area to be irrigated in hectares (acres), D is the gross depth of water applied in millimeters per hectare (inches per acre), F is the time of one complete irrigation cycle in days, and H is the number of hours the system operates per day.

The F and H factors are very important as these two factors have a direct bearing on the capital investment per hectare (acre) required for equipment. The greater the product of these two factors, the smaller the system capacity and the lower the equipment costs per hectare (acre) will be.

Design areas having different values of water replacement due to soil type, depth, or crop water peak use can be subdivided on the basis of water needed at the same time at each irrigation and the totals for each area added to give total system capacity needed.

WATER APPLICATION RATES

The rate (I) at which water should be applied to the soil by a sprinkler system depends upon the time required for the soil to absorb the calculated depth of water application without runoff. The depth of application (d) divided by this time of application gives the maximum application rate. The minimum application rate is that which will result in a uniform water distribution and high water use efficiency under the climatic conditions of the site. The selected application rate must be between these minimum and maximum values and most nearly fit the farm operation schedules.

Hand-move laterals, side-roll wheel move laterals, drag-type end tow laterals, pull-wheel end tow laterals, carriage with trailer lines, rotating boom, large volume gun, and solid set sprinkler systems all have their sprinklers stationary when actually applying water. Straight self-propelled laterals, circular self-propelled laterals, continuously moving boom, and continuously moving sprinkler machines have their sprinklers moving over the area while actually applying water.

SELECTION OF SPRINKLER

The rate of water application for the sprinklers that were on the stationary operating sprinkler systems is determined by the sprinkler head type, nozzle size, nozzle pressure, and spacing of sprinklers on the lateral and lateral spacing on the main pipeline. Each type of sprinkler head has certain water distribution pattern characteristics that change with nozzle sizes and operating pressures. Each sprinkler head has an optimum range of operating pressures for each nozzle size. When the pressure is higher than optimum, the water drops are smaller, fall closer to the sprinkler, and are more subject to pattern distortion due to wind. When the pressure is lower than optimum, the water drops are larger, and when the pressure is too low, the water falls in rings in the wetted area at a distance from the sprinkler head. The sprinkler head manufacturers recommend operating pressures or range of pressures, volume of water discharged at that pressure, and the wetted diameter for each size nozzle or nozzles and type of sprinkler head that will result in the most desirable application pattern.

Moisture Distribution Patterns and Sprinkler Spacing

The water depth distribution pattern for a single sprinkler operating under favorable conditions will be a triangular pattern with the largest depth of water at the sprinkler head and reducing to zero depth at the outer end of the wetted area. Because of this water depth distribution pattern to get as uniform a water distribution over the field as possible, it is necessary to overlap the sprinkler patterns.

When determining the sprinkler spacing on a lateral and lateral spacing on the main pipe line, general rules, where field tests have not shown otherwise, are sprinkler spacing on a lateral (S_L) should not exceed 50% of the wetted diameter given by the manufacturer for that sprinkler head nozzle size and operating pressure. Spacing of laterals on the main pipe line (S_m) should not exceed 65% of the wetted diameter. Where winds of 8 to 16 km (5 to 10 mi/hr) velocity are expected, the lateral spacing (S_m) on the main pipe line should be reduced to 50% of the sprinkler wetted diameter. Where winds over 16 km (10 mi/hr) are expected, the lateral spacing on the main line (S_m) should be limited to 30% of the sprinkler wetted diameter.

Sprinkler Discharge Requirements

The volume of water to be discharged by the individual sprinkler is a function of the application rate, previously determined, the spacing of the sprinklers on the lateral (S_L), and the spacing of the lateral pipe lines on the main pipe lines (S_m). A formula for calculating the required sprinkler discharge is:

$$q = \frac{S_L \times S_m \times I}{K} \qquad (5)$$

where q is the required sprinkler discharge in liters (gallons) per minute, S_L is the distance between sprinklers on the sprinkler lateral in meters (feet), S_m is the distance between lateral moves on main pipelines in meters (feet), I is the water application rate in millimeters (inches), and K is a constant 60.01 for metric and 96.25 for U.S.

For this sprinkler discharge, a sprinkler type, its nozzle size or sizes, optimum operating pressure, and diameter of wetted circle is selected from the manufacturer's performance tables. The assumed sprinkler spacing is then compared with the sprinkler wetted diameter to assure that the sprinkler meets the required spacing criteria. If it does not, a different nozzle size and operating pressure combination is selected and the checking procedure repeated.

SYSTEM LAYOUT

Having determined the area to be irrigated, the water depth to be applied, the frequency of irrigation, the sprinkler size, pressure, and spacing on the lateral pipe line, and the lateral pipe line spacing on the main pipe line, a preliminary sprinkler system is mapped out.

Using the scale map, layout this sprinkler system with the main pipe lines, lateral pipe lines, and sprinklers spaced as previously determined. The sprinkler system layout must provide for the simultaneous operation of the average number of sprinklers that will match the calculated sprinkler system capacity when operated at the design pressure. The average number of sprinklers operating simultaneously is determined by the following formula:

$$N = \frac{Q}{q} \qquad (6)$$

where N is the number of sprinklers operating simultaneously, Q is the total system capacity in liters (gallons) per minute, and q is the sprinkler head discharge in liters (gallons) per minute.

The variation in the number of sprinklers operating from time to time during an irrigation should be kept to a minimum to facilitate lateral routing and to maintain a near constant load on the pumping plant or gravity water supply. With a rectangular irrigation area, there should be very little variation in the number of sprinklers operating. On odd shaped fields, it is sometimes necessary to operate less than the average required number of sprinklers for one or more lateral settings. For many odd shaped tracts, the number of sprinklers will exceed the theoretical average number computed and allowances must be made for extra equipment to irrigate these parts of the tract.

The number of settings required to each lateral (operating at the same time with other laterals where more than one is involved) depends upon the number of allowable sets per day and the maximum number of days allowed for completing one irrigation during the peak use period. The required number of settings per lateral must not exceed the product of these two factors.

If the system layout provides for the theoretical number of sprinklers required, then the number of settings required per lateral will not exceed this allowable limit. Long narrow or irregular shaped parts of a tract may require additional lateral settings. Thus more equipment is needed if these areas are to be served within the allowable irrigation time period.

The lateral lines must be so located and must be of a pipe size and length that will result in a minimum variation in individual sprinkler discharges along the lateral line. The discharge variation should not exceed 10%. Lateral lines must be located and pipe sizes selected so the total losses due to pipe friction and elevation head will not exceed 20% of the designed operating pressure for the sprinklers. If possible, lateral pipe lines should be laid out as nearly on the level as possible. Laid down fairly uniform slopes, they will offset the pipe line friction loss and allow longer laterals. They should be laid at an angle to the prevailing winds. Running lateral pipe lines uphill should be avoided where possible, but if necessary, it will severely shorten the lateral to maintain uniform water distribution. Farming operations often influence the layout of lateral pipe lines and should be considered.

Main Line Layout

Main pipe lines or submains convey the water from the pump or other source to the lateral pipe lines in the area to be irrigated. The layout of these pipe lines should consider the area shape, topography, the scheduled operation of the laterals, the power cost and location of the water source for the lowest operating costs and capital investment.

Location of Water Source and Pumping Plants

The location of the water source near the center of the design area results in the least capital investment for main pipe lines and usually minimum pumping costs. If the source of the water is outside the area to be irrigated, it may be possible to locate the pump near the center of the area and bring water to the pump by gravity.

Booster pumps should be considered for small parts of a design area to avoid pumping all the water at high pressure when only a small area needs the high pressure. This saves energy.

ADJUSTMENTS TO MEET LAYOUT CONDITIONS

After completing the layout of main pipe lines and lateral pipe lines on the map, it may be necessary to make some adjustments in the number of operating sprinklers, water application rate, sprinkler discharge, operating time per day, frequency of irrigation, or total system capacity. This could result in a change in pipe sizes, sprinkler pressure, pump capacity, and power requirements to achieve a system that will be most efficient and economical for irrigating the area and crops.

After making the final adjustments in the layout and determining the lengths of the laterals, number of sprinklers supplied by each lateral, the volume of water carried by each lateral, the volume of water carried by the main and submain pipe lines, the pipe line sizes and friction losses along with the operating pressure at various vital locations in the system are calculated.

PIPE LINE DESIGN

Lateral pipe line sizes should be chosen so that the pressure variation in the pipe line due both to friction head loss and elevation head loss or gain does not exceed 20% of the design operating pressure head of the sprinkler lateral.

Main line or submain pipe sizes and materials should be chosen that will result in a reasonable balance between annual pumping costs and capital cost of the pipe line.

The loss in pressure caused by friction is the primary consideration in the design of any pipe system. Friction losses in pipe lines increase as the roughness of the inside surface of the pipe, as the length of the pipe line, as the velocity and viscosity of the water carried increases. Friction losses decrease as the pipe diameter and water density increases. The effects of viscosity and density have been neglected in pipe friction loss formulas developed for water flow in its natural state.

There are many formulas published for calculating the friction losses sustained by water flowing in pipes. Some pipe and sprinkler manufacturers have published tables, graphs, and slide rule calculations to assist in the calculation of friction loss in main pipe lines and sprinkler laterals.

Lateral Pipe Lines

More friction loss will be caused by the flow of water through the entire length of a lateral pipe line of given diameter than through the same lateral pipe line having a number of equally spaced operating sprinklers along its length. This is because the volume of water decreases by the amount of the sprinkler discharge as each sprinkler along the lateral is passed.

One method that has been used to calculate the friction loss in a sprinkler lateral pipe line is to start at the last sprinkler on the lateral and calculate the friction loss for the discharge of that sprinkler through the length of pipe back to the second from last sprinkler. The friction loss for each section of the lateral pipe line between sprinklers is then calculated, increasing the flow of water by the sprinkler discharge at the end of that section each time. A summary of the friction loss for each lateral section between sprinklers will give the total friction loss for the lateral.

The above method of calculating lateral friction loss requires many individual computations, which takes much time. Christensen developed a procedure in which the friction loss (H_f) is determined for a lateral pipe line carrying the total flow the full length of the lateral using the Scobey friction loss formula and multiplying this total loss by a factor (F) to obtain the lateral friction loss. This factor (F) was based on the number of sprinklers operating on the lateral and the velocity exponent used in Scobey's formula. Scobey's friction loss formula is:

$$H_f = \frac{K_s \times L \times V^{1.9} \times K}{1000 \times D^{1.1}} \tag{7}$$

where H_f is the friction loss in the lateral pipe line in meters per 30.48 m (feet per 100 feet) of pipe line, K_s is the coefficient of retardation based on the type and condition of the pipe material, L is the length of lateral in meters (feet), V is the velocity of flow in the lateral in meters per second (feet per second), D is the diameter of the pipe in meters (feet), and K is 0.79 for metric units and 1.0 for U.S. units. Christensen's formula for computing the factor (F) for multiplying the friction loss if the total water flowed the full lateral length to determine the actual lateral friction loss is

$$F = \frac{1}{m + 1} + \frac{1}{2N} + \frac{\sqrt{m - 1}}{6N^2} \tag{8}$$

where F is the factor (F), m is the velocity coefficient in Scobey's formula, and N is the number of sprinklers supplied water by the lateral pipe line.

Table 1 was prepared using Scobey's formula, K_s values, pipe diameters, and various volumes of water flow in the pipe to give the friction loss in 30.48 m (100 ft) length of lateral portable aluminum pipe in 9.14 m (30 ft) lengths between couplers.

Table 1
FRICTION LOSS IN METERS PER 30.48 METERS IN
LATERAL LINES OF PORTABLE ALUMINUM PIPE WITH
COUPLERS

(Based on Scobey's Formula and 9,14 Meter Pipe Lengths)[a]

Flow (ℓ/min)	50.8 mm[b] $K_s = 0.34$	76.4 mm[b] $K_s = 0.33$	101.6 mm[b] $K_s = 0.32$	127.0 mm[b] $K_s = 0.32$	152.4 mm $K_s = 0.32$
189.3	2.07	0.262	0.060		
227.1	2.95	0.369	0.085		
265.0	3.93	0.496	0.114	0.037	
302.8	5.09	0.640	0.148	0.048	
340.7	6.34	0.802	0.184	0.060	
378.5	7.74	0.975	0.225	0.073	0.030
454.2		1.38	0.317	0.103	0.043
529.9		1.86	0.426	0.138	0.057
605.6		2.39	0.549	0.180	0.074
681.3		2.99	0.689	0.223	0.092
757.0		3.66	0.841	0.273	0.113
832.7		4.39	1.01	0.326	0.132
908.4		5.15	1.19	0.384	0.159
984.1		6.00	1.38	0.448	0.185
1059.8		6.95	1.59	0.518	0.213
1135.5		7.89	1.82	0.588	0.243
1211.2		8.93	2.05	0.664	0.276
1286.9		10.00	2.30	0.747	0.311
1362.6		11.15	2.56	0.835	0.344
1438.3		12.37	2.85	0.924	0.384
1514.0		13.62	3.14	1.02	0.421
1589.7			3.44	1.12	0.460
1665.4			3.75	1.22	0.506
1741.4			4.08	1.33	0.549
1816.8			4.45	1.44	0.594
1892.5			4.82	1.55	0.646
2081.8			5.76	1.87	0.768
2271.0			6.77	2.20	0.908
2460.3			7.89	2.56	1.05
2649.5			9.08	2.95	1.22
2838.8			10.30	3.35	1.38
3028.0				3.81	1.57
3217.3				4.27	1.76
3406.5				4.75	1.96
3595.8				5.27	2.18
3785.0				5.79	2.40

[a] For 6.19 m pipe lengths, increase values in table by 7%. For 12.2 m lengths, decrease values by 3%.

[b] Outside diameter.

Table 2 was prepared using Christiansen's equation to show values of the F factor for the different number of water outlets from the sprinkler lateral.

An example of the calculation of the lateral head loss due to friction (H_f) in a 396 m (1300 ft) of 102 mm (4 in.) diameter aluminum lateral pipe line with couplers every 6.1 m (20 ft) and having 32 sprinklers spaced 12.2 m (40 ft) apart each discharging 27.2 ℓ (7.2 gal) of water per minute for a total lateral discharge of 869.6 ℓ (230 gal) per minute is:

Table 2
FACTOR (F), COMPUTED BY CHRISTIANSEN'S EQUATION, FOR USE IN COMPUTING FRICTION LOSS IN A LATERAL PIPE LINE WITH MULTIPLE SPRINKLER OUTLETS

Outlets (no.)	Value of F	Outlets (no.)	Value of F
1	1.000	22	0.368
2	0.634	23	0.367
3	0.528	24	0.366
4	0.480	25	0.365
5	0.451	26	0.364
6	0.433	27	0.364
7	0.419	28	0.363
8	0.410	29	0.363
9	0.402	30	0.362
10	0.396	31	0.361
11	0.392	32	0.360
12	0.388	33	0.360
13	0.384	34	0.360
14	0.381	35	0.359
15	0.379	40	0.358
16	0.377	50	0.355
17	0.375	60	0.353
18	0.373	70	0.352
19	0.372	80	0.351
20	0.370	90	0.351
21	0.369	100	0.350

$$H_f = \frac{396}{30.48} \times 1.10 \times 1.07 \times 0.361 = 5.52 \text{ m} = 54.1 \text{ kPa}$$

Obtain 1.10 from Table 1, 1.07 from footnote Table 1, and 0.361 from Table 2. The head loss in meters is converted to pressure loss in kilo-Pascals by multiplying by 9.806.

Where laterals are on level ground, the allowable pressure loss due to friction in the lateral pipe line should not be more than 20% of the average design lateral operating pressure for the type of sprinklers used (P_a).

The pressure requirements at the main pipe line (P_m) are determined by the formula:

$$P_m = P_a + zm = P_a + 3/4 \ P_f + P_r \tag{9}$$

where P_m is the pressure at the main pipe line inlet to the sprinkler lateral pipe line in kilo-Pascals (lb/in.^{2}), P_f is the friction loss pressure in the lateral in kPa (lb/in.^{2}), P_r is the pressure needed for the riser pipe in kPa (lb/in.^{2}); meters of water are converted to kPa by multiplying by 9.806, P_a is the average sprinkler design operating pressure in kPa (lb/in.^{2}), and the factor 3/4 is used to give the average operating pressure at the center rather than the end of the lateral.

An example of the calculation at the main pipe line (P_m) for the lateral used in the previous example, which has a designed sprinkler pressure (P_a) of 344.8 kPa (50 lb/in.^{2}) with a riser height of 1.0 m (3.28 ft), is

$$P_m = 344.8 + 3/4 \ 54.1 + 9.8 = 395.2 \text{ kPa}$$

Where laterals are laid uphill, the pressure loss due to the ground elevation gain (P_e) between the lateral inlet and last sprinkler on the lateral is subtracted from the allowed 0.20 P_a friction loss in the design of that lateral. The equation for determining the pressure required at the lateral main-line connection is

$$P_m = P_a + 3/4(P_f + P_e) + P_r \text{ kPa} \tag{10}$$

Where the sprinkler laterals are laid downhill, the pressure required at the main-line connection to lateral is calculated using the equation:

$$P_m = P_a + 3/4(P_f - P_e) + P_r \text{ in kPa} \tag{11}$$

Flow control valves or flow control sprinkler head nozzles are used on lateral pipe lines where the topography is too steep to limit pressure variation along the lateral pipe line to within 20% by selecting pipe sizes. Flow control valves have a friction loss (P_{cv}) that must be provided for in determining the pressure at the main pipe line (P_m). Flow control sprinkler head nozzles do not have the friction loss (P_{ev}). So the pressure requirements at the main pipe line are

$$P_m = P_a + P_f + P_e + P_{ev} \text{ kPa} \tag{12}$$

Main-Line Design

Sprinkler system main lines convey the volume of water required to operate the sprinkler laterals at the required sprinkler operating pressure from the pump or gravity source of water. The loss of pressure caused by friction and gain in elevation with the selection of economical pipe sizes and lengths are the design factors that must be considered.

Lateral operating schedules are important in main-line design where more than one lateral is operating at the same time. Sprinkler lateral settings can be scheduled so as to keep friction loss in the main lines at a minimum at all times. The operator must completely understand and follow the scheduling plan in operation to accomplish this.

When pumping supplies the system's water, the selection of pipe sizes and pipe materials is determined by a balance between pumping costs and the capitalized cost of the pipe selected. Where sprinkler pressure is furnished by gravity, the designer must select sizes that will conserve pressure.

On steep slopes, the design problem is one of selecting pipe sizes to dissipate excess gravity pressure by friction.

There are many types of main-line pipe materials such as steel, aluminum, asbestos-cement, and plastic. The manufacturer of each type of pipe line usually has friction loss tables for various classes of pipe. The designer should obtain copies of them for use.

The friction loss in many main pipe lines is calculated using one of three formulas (1) Scobey's, (2) Hazen-Williams', or (3) Manning's. Scobey's is generally used with pipe factor K_s equal to 0.32 for asbestos-cement pipe, 0.40 for aluminum pipe, and 0.36 for welded steel pipe 15 years old. Table 3 shows the friction loss per 30.48-m (100-ft) lengths of various size steel pipes and water flows.

Where only one lateral is moved along one or both sides of the main line, the pipe size is selected from the friction loss table for the type of main pipe line being considered for the liters (gallons) per minute input to the lateral.

Split lateral pipe-line layout consists of two or more laterals rotated around the main or submain pipe line. The purpose of the two-lateral operation is to equalize the load on the pump, regardless of the lateral position, and to move the laterals to their beginning points as part of the irrigation procedure. From the design scale map, the maximum flow of water

Table 3

FRICTION LOSS IN METERS PER 30.48 M IN MAIN LINES OF WELDED STEEL PIPE 15 YEARS OLD

(Based on Scobey's formula with K_s being 0.36)

Flow (ℓ/min)	101.6 — mm[a]			152.4 — mm[a]			203.2 — mm[a]			254.0 — mm[a]		
	16 gage	14 gage	12 gage	14 gage	12 gage	10 gage	14 gage	12 gage	10 gage	14 gage	12 gage	10 gage
189.3	0.066	0.067	0.075									
227.1	0.091	0.095	0.103									
265.0	0.120	0.125	0.135									
302.8	0.170	0.176	0.191									
340.7	0.206	0.216	0.234									
378.5	0.246	0.258	0.280	0.033	0.035	0.037						
473.1	0.390	0.408	0.442	0.052	0.055	0.059						
567.8	0.536	0.558	0.607	0.072	0.076	0.081						
662.4	0.738	0.768	0.832	0.098	0.104	0.110	0.023	0.024	0.026			
757.0	0.927	0.966	1.049	0.124	0.131	0.139	0.029	0.030	0.033			
946.3	1.46	1.53	1.66	0.196	0.207	0.220	0.046	0.048	0.051	0.015	0.016	0.016
1,135.5	2.05	2.15	2.33	0.275	0.290	0.309	0.066	0.068	0.072	0.021	0.022	0.022
1,324.8	2.74	2.87	3.11	0.369	0.387	0.413	0.087	0.091	0.095	0.029	0.030	0.031
1,514.0	3.51	3.69	3.99	0.472	0.497	0.530	0.112	0.116	0.122	0.037	0.038	0.039
1,703.3	4.39	4.66	4.97	0.591	0.622	0.662	0.140	0.145	0.152	0.046	0.047	0.049
1,892.5	5.36	5.67	6.07	0.719	0.759	0.807	0.170	0.177	0.186	0.056	0.058	0.060
2,271.0				1.033	1.088	1.158	0.244	0.254	0.266	0.080	0.083	0.086
2,649.5				1.375	1.451	1.542	0.326	0.338	0.355	0.107	0.110	0.115
3,028.0				1.765	1.859	1.979	0.418	0.433	0.455	0.137	0.141	0.147
3,406.5				2.219	2.340	2.488	0.527	0.546	0.573	0.175	0.178	0.185
3,785.0				2.713	2.859	3.012	0.643	0.664	0.699	0.212	0.217	0.225
4,542.0				3.810	4.023	4.241	0.927	0.935	0.980	0.298	0.305	0.318
5,299.0							1.213	1.262	1.319	0.396	0.411	0.428
6,056.0							1.564	1.625	1.705	0.512	0.530	0.551
6,813.0							1.960	2.030	2.123	0.646	0.661	0.689
7,570.0							2.396	2.473	2.607	0.799	0.808	0.842
9,462.5							3.658	3.810	3.979	1.210	1.234	1.285
11,355.0										1.682	1.750	1.827

[a] Outside diameters.

is determined from the movement of the sprinkler laterals. Using these maximum flow points on the main lines and the friction loss table for this type of main-line pipe, select a pipe size that gives a reasonable value of friction pressure loss. Adjustments are then made to balance annual pumping and capitalized pipe costs.

Where more than two laterals are operated and the flow in the main line is split, part being taken out at the first lateral and the rest continuing in the main line to serve other laterals, the design problem becomes more complex. No simple mathematical computations can be used to determine the minimum pipe sizes. Approximate sizes can be made by inspection, with trial and error calculations using the scale map of the system that shows the maximum flow points on the main and submain pipe lines.

The main line should be first designed to satisfy the requirements for operation with one lateral at the far end of the main line. It must be checked to see that it will satisfy the requirements for operation with the laterals in other positions on the main line. If the design does not satisfy the requirements for all operating conditions, adjustments in pipe size will have to be made until it does.

An example of main pipe line selection is for a sprinkler system on level land to irrigate an area 671 m (2200 ft) long and 792 m (2600 ft) wide, applying 101.6 mm (4.0 in.) of water every 11 days in the maximum crop use period at a 70% irrigation efficiency. This requires four laterals, each the same as shown in the lateral friction loss calculation section. These laterals are operated 23 hr and moved two times each day. The mainline discharges 871 ℓ (230 gal)/min per lateral. The system is operated with two laterals on each side of the 671 m (2200 ft) main pipe line. The water input to the main pipe line will be 3482 ℓ (920 gal)/min.

Using a welded steel main line and beginning at the section carrying 871 ℓ (230 gal)/min, this is from the 0 to the 168 m (0 to 550 ft) point, the friction is calculated. The next section is from the 168 to the 335 m (550 to 1100 ft) and will be carrying 1741 ℓ (460 gal)/min. The 335 to 503 (1100 to 1650 ft) section will be carrying 2612 ℓ (690 gal)/min. The 503 to 671 m (1650 to 2200 ft) section will be carrying 3482 ℓ (920 gal)/min.

Friction losses in these sections of main line are calculated using Table 3. For the first section listed above, enter the Table 3 at 871 ℓ, but since there is only 757.0 and 946.3 ℓ listed, interpolation will have to be made. So 946.3 − 757.0 equals 189.3, and 871 − 757 equals 114, the factor to multiply the friction loss difference between the two listed flows is 114 divided by 189.3, which equals 60%. Using a 12 gage 152.4 mm (6 in.) pipe line for the two listed flows, friction loss is 0.131 and 0.207 m. So the actual friction loss per 30.48 m length of pipe is (0.207 − 0.131) 0.60 + 0.131 = 0.177 m. Since there are 168 m in this section of pipe, the friction loss is (168 ÷ 30.48) 0.177 equals 0.976 m (3.21 ft).

Using this method, the friction loss for the other three sections are calculated. They are 168 to 335 m section 3.632 m (11.92 ft), 335 to 503 m section 1.819 m (5.54 ft), 503 to 671 m section 1.025 m (3.36 ft) for a total main pipe-line friction loss of 7.452 m (24.08 ft).

After a satisfactory main-line design has been completed for a total allowable friction loss, similar designs using other values of head loss can be used in balancing pipe and power costs in pumping.

PUMP REQUIREMENTS

The volume of water to be pumped and the total lift from the water source, friction losses, sprinkler pressure, and elevation changes in pascals (lb/in.^{2}) or in meters (ft) of head must be determined. Where operating conditions vary with the movement of lateral pipe lines or there is a change in the number of sprinklers operating, both maximum and minimum volumes of water and total lift must be determined. After determining the range of operating conditions, the pump and power unit can be selected with aid from pump manufacturers.

DESIGN COMPLETION

A map of the final system design should be prepared, showing the design area or areas to be irrigated, location of water supply and pumping plants with water supply pipe lines or canals, main and submain pipe lines, location of all sprinkler laterals and their direction of movement, spacing of sprinklers, pipe sizes and lengths in the lateral. A complete list of materials with their specifications should be furnished. This includes the sprinkler type with nozzle size or sizes, operating pressure, wetted diameter, and volume of water discharged. The main pipe-line pipe sizes and lengths, the pump discharge volume at the designed dynamic head, and horsepower requirements of the power unit should also be included.

REFERENCES

1. Soil Conservation Service, National Irrigation Handbook, Section 15, U.S. Department of Agriculture, Washington, D.C., 1968, chap. 11.
2. **Christansen, J. E.,** Irrigation by Sprinkling, Bull. 670, University of California, Berkeley, 1942.
3. **Pair, C. H., Hinz, W. W., Reid, C., and Frost, K. R.,** Sprinkler Irrigation, Sprinkler Irrigation Association, Silver Springs, Md., 1975.
4. **ASAE,** Use of SI (Metric) Units, ASAE EP 285.4, *Agricultural Engineers' Yearbook,* American Society of Agricultural Engineers, St. Joseph, Mich., 1979.

IRRIGATION EFFICIENCIES*

Terry A. Howell

INTRODUCTION

"Irrigation efficiency" is a widely used term[1-3] that expresses the performance index of irrigation systems or components of irrigation systems. Numerous attempts have been made to unify terminology on irrigation efficiency, and standards[4] have been proposed for this purpose, but the terminology is not unified. Several explanations can be offered to illustrate why unification is difficult. First, irrigation can have many purposes, and even a single irrigation often has multiple purposes. Second, the scale both in terms of time (a single irrigation, an irrigation season, etc.) and space (plot, field, farm, district, etc.) must be considered. Third, the irrigation method and irrigation management must be considered.

Irrigation efficiency as used here refers to the ratio, usually expressed in percent, of the irrigation water requirement (normally evapotranspiration plus leaching minus effective precipitation and soil moisture depletion) to the total amount of water diverted, stored, or pumped for irrigation. The irrigation water requirement could include any water used for "beneficial" or necessary purposes. The maximum possible efficiency is 100%. This definition of "irrigation efficiency" follows that proposed by Jensen,[5] but it broadens the concept to include other necessary water applications in the category with leaching and separates inputs of stored soil water and effective precipitation. The proper characterization of the "necessary" applications is disputable, and even the concept of leaching requirement has recently been reevaluated.[6] A theoretical irrigation system operating at 100% irrigation efficiency should be able to sustain permanent irrigated agriculture if no drainage impediments existed.

Irrigation systems comprise a wide range in physical designs, which require that "irrigation efficiency" be segmented to evaluate the entire system. Major segments of all irrigation systems are off-farm components and on-farm components. Both of these segments could have storage and conveyance facilities. Even though the physical configuration of off-farm and on-farm segments might be quite similar, a distinction usually exists for the segment management and control. The on-farm portion of the system is managed by a farm manager, while the off-farm portion of the system is managed by a district manager. The irrigation system operates through the control, management, legal and institutional arrangements of both segments. Not all irrigation systems have an off-farm segment since the water supply could be wholly developed on-farm, but many on-farm developed water supply irrigation systems are still subject to off-farm legal control through various national, state, and local regulations.

The off-farm and on-farm segments could consist of storage and conveyance components, which can be evaluated separately. The storage components could be any storage reservoir, either surface or subsurface, used for long-term or temporary storage or flow rate regulation. This storage component does not pertain to root zone soil moisture storage. The conveyance components could be any open or closed conveyance system used to transport water from an off-farm supply to the farm or an on-farm supply to the field. The on-farm segment has the water application component. This component consists of the application system to apply the required irrigation water to the field.

The relationship of segment and component efficiency terms is described below:

$$E_i = (E_{oi}/100)(E_{fi}/100)\ 100 \qquad (1)$$

* Contribution from the U.S. Department of Agriculture, Agricultural Research.

where E_i = irrigation efficiency (percent), E_{oi} = off-farm irrigation efficiency (percent), and E_{fi} = on-farm irrigation efficiency (percent).

$$E_i = [(W_{et} + W_{sp} - R_e - W_{sm})/W_i] \; 100 \tag{2}$$

where W_{et} = volume of evapotranspiration water per unit irrigated land area (depth of water), W_{sp} = volume of water applied per unit land area (depth of water) for special purposes, such as leaching, R_e = effective precipitation (precipitation minus runoff), W_{sm} = soil moisture depletion contributed to W_{et}, and W_i = volume of water per unit land area (depth of water) stored in a reservoir or diverted for irrigation.

$$E_{oi} = (E_{os}/100)(E_{oc}/100) \; 100 \tag{3}$$

where E_{os} = off-farm storage efficiency (percent), and E_{oc} = off-farm conveyance efficiency (percent).

$$E_{fi} = (E_{fs}/100)(E_{fc}/100)(E_{fa}/100) \; 100 \tag{4}$$

where E_{fs} = on-farm storage efficiency (percent), E_{fc} = on-farm conveyance efficiency (percent), and E_{fa} = field application efficiency (percent).

$$E_{fa} = [(W_{etf} + W_{spf} - R_{ef} - W_{smf})/W_a] \; 100 \tag{5}$$

where W_{etf} = volume of evapotranspiration water per unit land area (depth of water) of the field, W_{spf} = volume of water for special purposes applied per unit land area (depth of water) of the field, R_{ef} = effective precipitation received by the field, W_{smf} = soil moisture depletion in the field contributed to W_{et}, and W_a = volume of water applied per unit land area (depth of water) of the field.

The storage efficiencies — E_{os} and E_{fs} — are defined as the ratio of the volume of water per unit land area (depth of water) delivered from the reservoir for irrigation to the volume of water per unit land area (depth of water) delivered to the reservoir for irrigation. On-farm storage should include any tailwater pits used to recover irrigation runoff. The conveyance efficiencies — E_{oc} and E_{fc} — are characterized as the ratio of the volume of water per unit land area (depth of water) delivered by the conveyance system for irrigation to the volume of water per unit land area (depth of water) delivered to the conveyance system for irrigation.

Many of these efficiency parameters are dynamic; therefore, for clarity and comparative purposes, many complementary parameters such as irrigation unit size, field size, time periods, number of irrigation measurement procedures, computational procedures, types of irrigation systems, crop performance, etc. should be specified.

In many cases, the uniformity of irrigation is important. Parameters such as the uniformity coefficient[7] and distribution efficiency[8] have been used as indexes of application uniformity. Although neither of these terms is an actual efficiency, they directly affect irrigation efficiency. The uniformity coefficient (C_u) is given by

$$C_u = [1.0 - (\Sigma x/Mn)] \; 100 \tag{6}$$

where x = absolute values of the deviation of individual observations from the mean, M = the mean of the observations, and n = the number of observations. Although originally intended for sprinkler irrigation systems, the uniformity coefficient can also be applied to surface irrigation and trickle irrigation systems. Hart[9] showed that $C_u = [1.0 - (2/\pi)^{1/2}C_v]100$

if the irrigation distribution is a normal distribution where C_v = the coefficient of variation (standard deviation divided by the mean).

WATER CONSUMPTION BY IRRIGATION

Water diverted for irrigation is used either consumptively or nonconsumptively. Water is consumed when it is evaporated or transpired into the atmosphere where it can only be recovered as condensation or precipitation in the hydrologic cycle. Crop production (biomass) increases in direct proportion to transpiration for many plant species. Soil water evaporation, reservoir evaporation, canal evaporation, spray evaporation, and phreatophyte transpiration are nonbeneficial consumptive uses. Nonconsumptive uses of irrigation water are percolation (seepage), runoff, and operational uses (channel detention, spills, etc.). Nonconsumptive uses, except those that enter nonrecoverable ground water, can usually be reused downstream.

The magnitude and recovery of the nonconsumptive water varies widely. Jensen[10] estimated the effective irrigation efficiency as follows:

$$E_e = E_i + E_r(100 - E_i) \tag{7}$$

where E_e = effective irrigation efficiency (percent), and E_r = fraction of nonconsumptive water use which is or can be recovered. Figure 1 illustrates an irrigation water budget for the 17 western states in a "normal" water year with 1975 level of development. The irrigation efficiency is 41% but the effective irrigation efficiency is 87%. Therefore, even when significant quantities of water are inefficiently used, a significant portion of this water may be recovered either as ground water or downstream flow.

The inefficiency of irrigation creates several undesirable situations. Even though significant quantities of water may be available for reuse, the quality of that water is often markedly reduced by increased salinity. In addition, energy must be expended to recover the ground water. In some areas, excess percolation from overirrigation leads to shallow perched water tables that reduce crop productivity and may require artificial drainage. Overirrigation with ground water is one cause of ground-water overdrafts, which can result in ground subsidence or saltwater intrusion into fresh-water aquifers. In addition, inefficiency of water use results in overdesign of irrigation systems, particularly storage and conveyance structures, and the use of excess energy to pump the water in certain situations.

MEASUREMENT OF IRRIGATION EFFICIENCY PARAMETERS

Irrigation efficiency and component efficiencies are evaluated by measurements of the irrigation system parameters necessary to compute the factors in the equations in Section on Water Consumption by Irrigation. The measured parameters are integrated over the time and the land areas appropriate to the desired analysis. Merriam and Keller[12] and Karmeli et al.[13] present several techniques for evaluating and predicting on-farm irrigation efficiencies.

The measurement method used to determine each parameter would depend on the type of irrigation system being studied. Regardless of the measurement method, water flow rate is extremely difficult to measure to accuracies exceeding plus or minus 5% for long time periods in open channels and plus or minus 2% in closed conduits. The measurement of evapotranspiration is difficult and subject to several errors. Water balance techniques are normally used so that evapotranspiration is determined indirectly from changes in soil moisture. Often evapotranspiration is estimated using various empirical procedures.[14] Most of these procedures only estimate the evapotranspiration by "adequately" watered crops. Limited water applications, which result in plant water stress, require rather advanced techniques for accurate estimation of evapotranspiration. Measurements and estimates of seasonal eva-

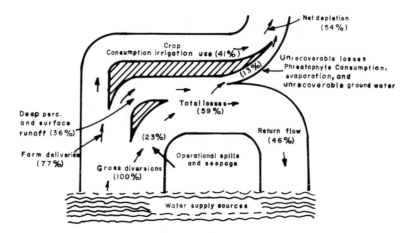

FIGURE 1.   Irrigation water budget for the 17 western states for a normal water supply year with 1975 level of development. (From Interagency Task Force Rep., Irrigation Water Use and Management, Environmental Protection Agency, U.S. Department of the Interior, U.S. Department of Agriculture, Washington, D.C., 1979, 23. With permission.)

potranspiration could contain errors of plus or minus 5%, and estimates for shorter time periods could have larger errors. Combination of the errors in flow measurement and evapotranspiration could conservatively result in errors of plus or minus 8%, or more, in determining efficiency components.

Errors in design resulting from incorrect estimates of irrigation efficiencies in some cases have had disasterous consequences in the operation of irrigation projects. Overestimating irrigation efficiency has resulted in the inclusion of too much land in the project. This has led to underirrigation of part of the project and has reduced the crop production potential. Underestimating irrigation efficiency, in contrast, has resulted in investment of excess capital and smaller projects than required. The excess water supply leads to wasteful irrigation practices and usually to drainage problems. In some cases, inefficiency is perpetuated by legal and institutional arrangements that penalize improvements. Even privately developed irrigation systems can be penalized for improvements when energy charges are based on a declining rate basis (the cost per unit decreases as the quantity used increases).

IRRIGATION EFFICIENCIES

Irrigation efficiencies for the U.S. were estimated by the U.S. Department of Agriculture Soil Conservation Service[15] for the Second National Water Assessment.* Table 1 summarizes primary irrigation characteristics in the 17 western states. The average off-farm conveyance efficiency for the U.S. was estimated at 78% and the on-farm efficiency at 53%, giving an overall irrigation efficiency of 41%. Jensen[16] recently summarized many measurements of irrigation efficiency that have been reported in the literature. Table 2 presents his summary along with his estimates of attainable efficiencies for field and farm systems. Bos and Nugteren[17] reported field application efficiency ranged from 40 to 75% (averaged 60%) for 32 irrigated areas including Australia and the U.S. Jensen[16] concluded that excessive water application during an irrigation was the greatest single factor contributing to low application efficiency.

The method of water application directly affects the on-farm application efficiency. The

---

*   The Nation's Water Resources — 1975-2000, Vol. 1, U.S. Water Resources Council, Washington, D.C., 1978.

Table 1A
IRRIGATION CHARACTERISTICS IN THE 17 WESTERN STATES

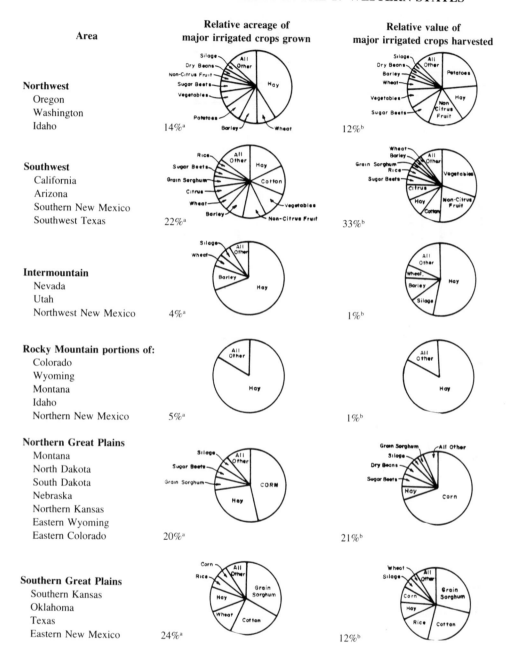

	Relative acreage of major irrigated crops grown	Relative value of major irrigated crops harvested
Area		
Northwest Oregon Washington Idaho	14%[a]	12%[b]
Southwest California Arizona Southern New Mexico Southwest Texas	22%[a]	33%[b]
Intermountain Nevada Utah Northwest New Mexico	4%[a]	1%[b]
Rocky Mountain portions of: Colorado Wyoming Montana Idaho Northern New Mexico	5%[a]	1%[b]
Northern Great Plains Montana North Dakota South Dakota Nebraska Northern Kansas Eastern Wyoming Eastern Colorado	20%[a]	21%[b]
Southern Great Plains Southern Kansas Oklahoma Texas Eastern New Mexico	24%[a]	12%[b]

[a] Total U.S. irrigated acreage.
[b] Total value of U.S. irrigated crops.

From Interagency Task Force Rep., Irrigation Water Use and Management, Environmental Protection Agency, U.S. Department of the Interior, U.S. Department of Agriculture, Washington, D.C., 1979, 23.

Table 1B
IRRIGATION CHARACTERISTICS IN THE SEVENTEEN WESTERN STATES

Area	Irrigation situation	Crop growing season (days)	Consumptive irrigation requirement for alfalfa[a] (m)	Water source	Irrigation methods	Present average irrigation efficiencies (%)		Competitive uses	Instream flow	Water quality
						On-farm	Off-farm conveyance			
Northwest	Intermediate valley	120—200	0.7—0.8	70% Surface 30% Ground water	About equally divided between sprinkler and surface (border, furrow, basin, corrugations), some trickle	25—70	60—95	Hydropower, recreation, instream flows, new irrigation and navigation	Seasonal inadequacies in streams and estuaries	Seasonal temperature fluctuations, dissolved gases and sediment
Oregon Washington Idaho	Mountain meadow	80—120	0.4—0.5	100% Surface	Wild flood	25—40	55—70	Recreation and instream flows	Inadequate streamflows in dry years	Excellent
Southwest	Lower valley[b]	200—365	0.9—1.9	80% Surface 20% Ground water	Surface (border, basin, corrugations, furrow); sprinkler and some trickle	50—70	70—95	M and $\bar{I}$, recreation, estuary inflow, instream flows, new irrigation	Severely depleted streams and estuary inflows	Salinity increases downstream
California	Intermediate valley	100—200	0.6—1.3	90% Surface 10% Ground water	Surface (border, contour ditch furrow and corrugations)	45—65	70—80	M and $\bar{I}$, hydropower, recreation and instream flows	Seasonal inadequacies	Good
Arizona Southern New Mexico Southwest Texas	Plains (with on-farm water supply)	200—365	0.9—1.9	100% Ground water (severe overdraft)	Surface (border, basin corrugations, furrow), some sprinkler	60—70	None	M and $\bar{I}$ and new irrigation	Inadequacies	Ground water good to poor
Intermountain	Intermediate valley[b]	120—200	0.6—1.1	80% Surface 20% Ground water	Surface (border, contour ditch furrow, corrugations), some sprinkler	35—50	70—95	Industry, recreation and instream flows	Seasonal inadequacies	Salinity increases downstream
Nevada Utah Northwest New Mexico	Mountain meadow	80—120	0.4—0.5	100% Surface	Wild flood	25—50	50—80	Recreation and instream flows	Localized areas with inadequate flows in dry years	Excellent
Rocky Mountain	Mountain meadow[b]	80—120	0.3—0.5	100% Surface	Wild flood	25—50	50—80	Recreation, instream flows and transbasin diversion	Localized areas with inadequate flows in dry years	Excellent

Portions of	Land type									
Colorado, Wyoming, Montana, Idaho, Northern New Mexico	Intermediate valley	100—150	0.6—0.8	95% Surface 5% Ground water	Surface (border, contour ditch, corrugations and furrow), some sprinkler	40—55	50—95	Power generation, recreation and instream flows	Seasonal inadequacies	Salinity increases downstream
Northern Great Plains	Plains (with on-farm water supply)[b]	160—240	0.5—0.8	90% Ground water 10% Surface (overdraft)	About equally divided between sprinkler and surface (furrow)	40—65	None	M and Ī and new irrigation	Seasonal inadequacies	Ground water good
Montana, North Dakota, South Dakota, Nebraska, Northern Kansas, Eastern Wyoming, Eastern Colorado	Intermediate valley	150—200	0.5—0.8	90% Surface 10% Ground water	Surface (border, contour ditch, corrugations and furrow), some sprinkler	40—55	40—90	Industry (power generation), recreation and instream flows	Seasonal inadequacies	Warm summer temperatures and sediment problems
Southern Great Plains	Plains (with on-farm water supply)[b]	180—330	0.4—1.4	95% Ground water (severe overdraft)	About equally divided between sprinkler and furrow and basin	50—70	None	M and Ī and new irrigation	Seasonal inadequacies	Ground water good to poor
Southern Kansas, Oklahoma, Texas, Eastern New Mexico	Lower valley	180—330	0.4—1.4	50% Surface 50% Ground water	Surface (furrow, border and basin), some sprinkler	65—75	80—95	M and Ī, estuary inflow, instream flows and recreation	Severely depleted streams and estuary inflows	Low to high salinity

[a] A widely grown crop — data indicate relative irrigation water requirements among areas. However, production per acre varies also.
[b] Indicates the major irrigated acreage in the area.

From Interagency Task Force Rep., Irrigation Water Use and Management, Environmental Protection Agency, U.S. Department of the Interior, U.S. Department of Agriculture, Washington, D.C., 1979, 23.

Table 2
FIELD WATER APPLICATION AND FARM IRRIGATION EFFICIENCIES[16]

Irrigation method	Field efficiency (%) Attainable E^*_{fa}	Observed (E_{fa}) Range	Observed (E_{fa}) Average	Farm efficiency (%) Attainable E^*_n	Observed (E_n) Range	Observed (E_n) Average	Comments
Surface							
Furrows and rills	70	—	58	65	—	—	Survey results, ICID
	70	—	—	65	31—55	42	5 Idaho farms, 5-year period
	70	—	—	65	29—41	35	5 Columbia Basin farms, 3-years
	70	57—73	65	65	—	—	Nebraska, experimental field
	70	24—77	46	—	—	—	Potatoes; Utah 1959
With auto, cutback	85	—	76	80	—	—	Experimental field
With runoff reuse	90	84—97	92	85	—	—	Nebraska, experimental field
Multiset	80	—	—	75	—	—	Experimental field
Multiset and reuse	90	—	—	85	—	—	Experimental field
Multiset and auto. reuse	95	—	95	90	—	—	Experimental field
Graded borders	80	—	57	75	—	—	Survey results, ICID
Low gradient and level borders	90	56—100	83	85	—	—	Experimental borders
Basins	75	—	59	70	—	—	Survey results, ICID
Basins (small)	75	25—40	30	65	—	—	Canal + public wells (Pakistan)
		0—100	40	65	—	—	Canal + private wells (Pakistan)
Basins (level)	85	40—100	60	70	—	—	Canal water only (Pakistan)
Basins (laser beam leveled)	90	85—100	—	80	—	—	
		—	—	85	—	—	
Sprinkler							
Standard set and side roll	80	—	68	75	—	—	Survey results
	80	—	—	75	33—62	—	Columbia basin, 2 farms
	80	61—65	—	75	—	—	Experimental fields
Center pivot	90	—	—	85	48—60	53	Columbia basin
	90	—	—	90	—	—	Eastern Colorado

Trickle

Row crops   95   —   Equivalent to best furrow systems

Note: Specific literature citations are deleted to satisfy space limitations. The reader is referred to Jensen, M. E., Irrigation Water Management for the Next Decade, Proc. New Zealand Irrig. Conf., Ashburton, 1978, 245, for specific details.[16]

primary irrigation methods include surface, sprinkler, and trickle irrigation (drip). Although each method has several advantages when compared with each other method, no one method is consistently more efficient that the others. The methods do differ in the range of potential efficiencies that could be obtained.[3,16]

Surface Irrigation

The main losses of water from surface irrigation are tailwater runoff, deep percolation, and direct evaporation. The amount of tailwater and deep percolation depend on the type of application system, management, and soil characteristics. Evaporation depends on the length of ponding and environmental factors.

On-farm irrigation efficiencies for surface irrigation systems can range from 60 to 80%[3] with uniformities exceeding 80%. Tailwater reuse can significantly improve on-farm irrigation efficiencies possibly to the 80 to 90% range.[18] Land leveling with lasers[19] also can improve irrigation efficiencies. Cut-back furrow stream irrigation can improve efficiencies, but it has not proved practical as yet. Automation and computer control of pumps, valves, and head gates may make cut-back irrigation possible in the future.

Sprinkler Irrigation

The main losses of water from sprinkler systems are runoff, spray drift, and evaporation. Evaporative losses can be as great as 30%[20] and depend on atmospheric conditions of wind speed and vapor-pressure deficit as well as sprinkler characteristics and management. Usually for a given sprinkler and nozzle, increased pressure results in increased spray atomization and evaporation. High wind speeds and large vapor-pressure deficits increase evaporation from sprinklers. Runoff losses from some center pivot and volume gun systems may approach 25%, but usually are less. Sprinkling soils having a very low intake rate may produce runoff from sloping land. Furrow diking has been effective in reducing runoff in these cases.

The on-farm irrigation efficiency of sprinkler systems can range from 60 to 85%.[20] Uniformity of distribution may range from 75 to 95% for many systems. The measurements of uniformity of application are many times inconsistent. Solomon[21] has presented information on the variability of Cu data. Figure 2 shows the number of tests required to measure Cu within 1% tolerance for 80, 90, and 95% statistical confidence limits. For other Cu tolerance limits, divide the number of trials from Figure 2 by the square of the Cu tolerance in percent (for example, with a ±5% tolerance, divide the number of trials by 25). Clearly, Cu cannot be specified accurately without conducting several tests. Hydraulic pressure variation, wind, and individual sprinkler characteristics reduce uniformity.

Trickle Irrigation

The main water losses from trickle systems are from deep percolation, runoff, and direct evaporation. The direct evaporation may be insignificant for widely spaced plants with fully developed plant cover, but could be several millimeters per day for closer plant spacings with developing covers. Runoff can occur when trickling on sloping soils that have a low intake rate.

The on-farm irrigation efficiency of trickle irrigation systems can range from 70 to 90%. Uniformity of application can range from 60 to 95%[22] for several large-scale systems. The sources of nonuniformity are hydraulic pressure variation, emitter manufacturing variation, emitter clogging, and thermally induced effects (expansion and viscosity effects). Solomon[23] illustrated the interaction of pressure variation with manufacturing variation. Emitter manufacturing variation in many cases had a larger impact on irrigation efficiency than friction loss (up to 50% of emitter pressure) or emitter flow exponent.

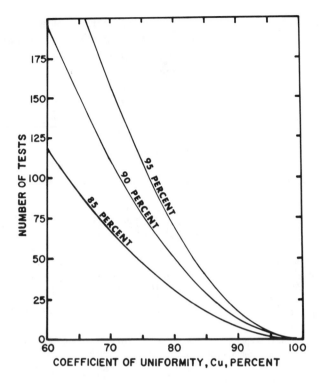

FIGURE 2.   Number of tests of sprinkler coefficient of uniformity required for specific statistical confidence levels for plus or minus 1% in Cu. (From Solomon, K., *Trans. ASAE*, 22, 1078, 1979. With permission.)

SUMMARY

Irrigation efficiency is dynamic and to a large extent depends on many parameters that cannot be easily controlled by mechanical devices. Irrigation efficiency can be improved through

. . . an accelerated program of improving irrigation systems and water management, using the best practical technology available; changes in water rights laws in some states to encourage conservation of water; an evaluation through soil surveys of the soils suitable for irrigation; state land-use planning that would encourage the most efficient agricultural lands to remain in agriculture; acceleration of assistance programs that emphasize the development and implementation of improved technology, irrigation systems and cultural practices; and a modified water-pricing policy that encourages good water management.[15]

Although a high irrigation efficiency is possible for many systems, crop productivity is more directly related to distribution uniformity. Most irrigation systems have low application efficiencies when the application is distributed to "adequately" irrigate all points in a field. Although this practice usually results in uniform crop growth and yield, excess water is applied. Underirrigating a portion of the field (for example 5 or 10%) could increase the on-farm application efficiency by 10% or more. The optimum operating point must be determined by a complete engineering, agronomic, and economic analysis.

There is a common misconception that improving irrigation efficiencies will automatically result in water conservation with the "extra" water available for supply. Irrigation efficiency can be increased by either increasing water consumption (holding water diverted constant) or by decreasing diverted water (holding consumption constant).[16] In either case, the amount conserved and available for competing users or uses is normally not in direct proportion to the increase in irrigation efficiency. Water savings determinations must be made on a hydrologic basin level and not the farm or district level. The hydrologic basin may encompass local, state, or even national boundaries. Additional justification for improving irrigation efficiency (in addition to water conservation) is improved overall water quality and energy savings from reduction in irrigation pumping.

REFERENCES

1. **Israelsen, O. W. and Hansen, V. E.**, *Irrigation Principles and Practices*, John Wiley & Sons, New York, 1962, chaps. 1 and 13.
2. **Jensen, M. E., Swarner, L. R., and Phelan, J. T.**, Improving irrigation efficiencies, in *Irrigation of Agricultural Lands*, Agronomy No. 11, Hagan, R. M., Haise, H. R., and Edminster, T. W., Eds., American Society of Agronomy, Madison, Wis., 1967, chap. 61.
3. **Willardson, L. S.**, Attainable irrigation efficiencies, *J. Irrig. Drain. Div. ASCE*, IR2, 239, 1972.
4. **Bos, M. G.**, Standards for irrigation efficiencies of ICID, *J. Irrig. Drain. Div. ASCE*, IR1, 37, 1979.
5. **Jensen, M. E.**, Evaluation irrigation efficiency, *J. Irrig. Drain. Div. ASCE*, IR1, 83, 1962.
6. **van Schilfgaarde, J., Bernstein, L., Rhoades, J. D., and Rawlins, S. L.**, Irrigation management for salt control, *J. Irrig. Drain. Div. ASCE*, IR3, 321, 1974.
7. **Christiansen, J. E.**, Irrigation by Sprinkling, Bull. 670, Calif. Agric. Exp. Stn., University of California, Berkeley, 1942, 94.
8. **Hansen, V. E.**, New concepts in irrigation efficiency, *Trans. ASAE*, 3, 55, 1960.
9. **Hart, W. E.**, Overhead irrigation pattern parameters, *Agric. Eng.*, 42, 354, 1961.
10. **Jensen, M. E.**, Water conservation and irrigation systems, in *Proc. Climate-Technology Seminar*, University of Missouri, Columbia, 1977, 209.
11. **Interagency Task Force Rep.**, Irrigation Water Use and Management, Environmental Protection Agency, U.S. Department of the Interior, U.S. Department of Agriculture, Washington, D.C., 1979, 23.
12. **Merriam, J. L. and Keller, J.**, *Farm Irrigation System Evaluation: A Guide for Management*, Utah State University, Logan, 1978, chap. 1—12.
13. **Karmeli, D., Salazar, L. J., and Walker, W. R.**, Assessing the Spatial Variability of Irrigation Water Applications, Environ. Protect. Ser. EPA-600/2-78-041, Environmental Protection Agency, Ada, Okla., 1978, chap. 1—6.
14. **Doorenbos, J. and Pruitt, W. O.**, Crop Water Requirements, FAO Irrig. and Drain., Paper 24, Food and Agriculture Organization, Rome, 1977.
15. Crop Consumptive Irrigation Requirements and Irrigation Efficiency Coefficients for the United States, Soil Conservation Service, U.S. Department of Agriculture, Washington, D.C., 1976.
16. **Jensen, M. E.**, Irrigation Methods and Efficiencies, paper presented at a World Bank Seminar, Washington, D.C., January 8, 1980, 41.
17. **Bos, M. G. and Nugteren, J.**, On Irrigation Efficiencies, 2nd ed., Publ. No. 19, Ant. Inst. for Land Reclamation and Improvement, 1978, 142.
18. **Linderman, C. L. and Stegman, E. C.**, Seasonal variation of hydraulic parameters and their influence upon surface irrigation application efficiency, *Trans. ASAE*, 14, 914, 1971.
19. **Dedrick, A. R., Replogle, J. A., and Erie, L. J.**, On-farm level-basin irrigation — saves water and energy, *Civ. Eng.*, 48(1), 60, 1978.
20. **Pair, C. H., Hinz, W. W., Reid, C., and Frost, K. R.**, *Sprinkler Irrigation*, 4th ed., Sprinkler Irrigation Association, Silver Springs, Md., 1975, chap. 5 and 7.
21. **Solomon, K.**, Variability of sprinkler coefficient of uniformity tests results, *Trans. ASAE*, 22, 1078, 1979.
22. **Bliesner, R. D.**, Field evaluation of trickle irrigation efficiency, in *Water Management for Irrigation and Management*, Vol. 1, American Society of Civil Engineers, New York, 1977, 382.
23. **Solomon, K.**, Manufacturing variation of trickle emitters, *Trans. ASAE*, 22, 1034, 1979.

VOLUME GUN SYSTEMS*

Hollis Shull

VOLUME GUN SPRINKLERS

Volume gun sprinklers are large, single nozzle, rotating sprinklers that operate with water pressures from 345 to 900 kPa (50 to 130 lb/in.^{2}) and discharge from 0.3 to more than 4.6 m^3/min (75 to 1200 gal/min). Nozzle diameters range from about 15 to 50 mm (0.6 to 2 in.). Two nozzle configurations, smooth bore (straight or tapered) and ring (orifice), are available. Smooth bore nozzles provide greater coverage diameter and less stream breakup than do ring nozzles under otherwise identical conditions. The sprinklers are impact rotated, have adjustable rotation speed, and are available with either full circle, or adjustable angle, part circle coverage. The diameter wetted by a single sprinkler ranges from about 60 m (200 ft) for smaller models to over 180 m (600 ft) for the larger ones. Sprinklers are available with trajectory angles from 18 to 32°.[1,2] The lower angles are usually used in windy areas.

Because high water pressure is required for sufficient stream breakup to prevent crop damage and to achieve uniform distribution, the pumping power requirement for volume gun sprinklers is high.

Volume gun sprinklers are poorly adapted to soils with low infiltration rates because the minimum average application rate is about 7.6 mm/hr (0.3 in./hr).[2] Like all impact-rotated sprinklers, it applies water intermittently and repetitively as it rotates. Because of the high discharge rate from volume gun sprinklers, average water application rates are high, and instantaneous rates are very high. Most soils need surface storage to hold the applied water in place until it has infiltrated.

Volume gun sprinklers can be operated as intermittently moved or as continuously moving (traveling) systems. With intermittently moved systems, the sprinkler is usually moved manually or mechanically in a square or rectangular pattern. Circular areas are irrigated. With traveling systems, the sprinkler is mounted on a mobile, wheeled carrier and is towed through the field at low speed along parallel travel lanes. Parallel strips are irrigated.

Intermittently moved, volume gun sprinklers are well suited to irrigating areas where soils have high infiltration rates, where low first cost is important, and high water pumping costs can be tolerated, where field obstructions make the use of other irrigation systems difficult or impossible, or where field boundaries are irregular.[3] Water can be supplied to the sprinkler through portable aluminum pipe laterals with valved outlets. The sprinkler will be moved periodically from outlet to outlet along the lateral. When it has moved the entire lateral length, the lateral is moved to the next location and the sprinkler moving sequence repeated. If enough pipe is available, periodic lateral line moving can be eliminated by placing all of the required laterals in the field at the beginning of the irrigation season. Buried pipe, with valved outlets rising above ground level at the desired sprinkler spacings, can also be used. A combination of buried pipes and portable laterals is another alternative. Sprinklers should be spaced about 60% of the wetted diameter for operation in low or no-wind areas.

Traveling volume gun systems are suited to the same conditions as intermittently moved systems.[4] Traveling systems require less labor than do intermittently moved systems, at higher initial cost. Traveling volume guns are well adapted to the disposal of waste water that contains solid materials because their large diameter nozzles are not easily clogged.

Traveling volume gun systems are towed down parallel travel lanes with a cable that winds on a rotating drum. The drum is powered either by a water motor, with the drum

---

*   Contribution from U.S. Department of Agriculture, Science and Education Administration, Agricultural Research.

located on the carrier, or by an internal combustion engine, with the engine and drum located either on the carrier or at the end of the travel lane.

Water is supplied to the traveling sprinkler through a flexible hose that is pulled behind the carrier. Because the length of hose dragged increases as the carrier moves through the travel lane, travel speed will decrease as distance traveled increases, resulting in nonuniform water application. An internal combustion engine with a constant speed governor will maintain a more uniform speed than will a water-driven motor.

Wind distorts the water-distribution pattern from volume gun sprinklers.[5] For intermittently moved systems, sprinklers must be spaced closer than under no-wind conditions to maintain satisfactory application uniformity. Under windy conditions, sprinklers should be spaced about 50% of the no-wind wetted diameter.[3] If rectangular spacing is used, the closer spacings should be perpendicular to the prevailing wind direction.[3] Staggering the sprinklers, to obtain triangular spacing, will further improve the distribution pattern.[3] Distribution patterns can also be improved by using sprinklers with lower trajectory angles; however, to obtain good stream breakup, low-trajectory-angle sprinklers should be operated at a water pressure about 70 kPa (10 lb/in.[2]) greater than would be used with a higher trajectory-angle sprinkler.[3]

When traveling gun sprinkler systems are operated in windy areas, the travel lane spacing must be reduced to obtain good water-distribution uniformity. If possible, travel lanes should be at right angles to the prevailing wind direction.

Research has shown that in windy areas, the selection of travel lane spacing is not so much that of determining the spacing for a specific wind speed, but of selecting a spacing to produce satisfactory uniformity under any and all wind speeds and directions that occur during the irrigation season.[6] The maximum lane spacing that will provide acceptable uniformity at a 3.5 m/sec (8 mi/hr) wind speed will provide an entirely unacceptable uniformity with low or no wind. A travel lane spacing narrower than necessary for one specific wind condition is needed to obtain acceptably uniform application under day-to-day wind variations in a windy area.

With either intermittently moved or traveling volume gun irrigation systems, wind increases the labor required to irrigate a given area because it increases the frequency of sprinkler moves or travel lane changes.

ROTARY-BOOM SPRINKLER SYSTEMS

Rotary-boom sprinkler systems resemble volume gun systems in their specifications and performance. A rotary boom has arms from 21 to 42 m (70 to 140 ft) in length extending from the center. The arms are fitted with nozzles that, because of their discharge angle, cause the boom to rotate. Rotary booms can be operated as either intermittently moved or traveling systems. Although their operating characteristics resemble those of the volume gun, their distribution pattern is somewhat less affected by wind. Because of their long boom arms, they are more difficult to maneuver in the field.

Wiersma,[7] who tested large sprinklers in 1958, determined that under windy conditions boom sprinkler spacing could be greater than volume gun spacing. However, in 1958, boom sprinklers were in the developmental stage and drag hoses had not yet been developed, so he tested only intermittently moved systems. He recommended that boom sprinklers not be operated in winds greater than 5.4 m/sec (12 mi/hr), and that volume gun sprinklers not be operated in winds greater than 4.0 m/sec (9 mi/hr). He also found that rotary-boom sprinklers will operate at lower pressure than will volume gun sprinklers to give equivalent droplet breakup.

Allred and Machmeier[8,9] conducted both laboratory and field studies of the performance of boom sprinklers. They found that wind caused the rotational boom speed to vary as the boom rotated. The rotational speed was maximum when the boom direction was parallel to the wind direction. They determined that a relatively fast rotational speed of approximately 1 rpm, reduced the rotational speed variation. With this fast rotational speed, they were able to reduce the water pressure to 310 kPa (45 lb/in.^{2}) and obtain acceptable water stream breakup.

DESIGN OF VOLUME GUN IRRIGATION SYSTEMS

To design volume gun irrigation systems, information is needed on the following factors:

1. Water supply
2. Water pressure available at the sprinkler
3. Sprinkler nozzle diameter
4. Size and shape of irrigated area
5. Wind conditions expected
6. Nozzle type (smooth bore or ring)
7. Available labor
8. Soil water-holding capacity
9. Crop water requirement
10. Crop rooting depth

The water supply and crop water requirement will determine the size of the area that can be irrigated.

The required nozzle size for the available water supply and pressure can be determined from manufacturer's literature. If no such literature is available, Equation 1 can be used to estimate the sprinkler discharge as a function of nozzle diameter and water pressure.

$$Q = CVA \qquad (1)$$

where Q is discharge, volume per unit time, C (ring nozzle) $= 0.01D + 0.326$, C (smooth bore nozzle) $= 0.99 - 9.43/P$, V is discharge velocity $= (2\ gH)^{1/2}$, g is gravitational acceleration, H is water pressure, A is nozzle cross-sectional area, P is operating pressure, kPa, and D is nozzle diameter, mm. For Q in m^3/sec, express V in m/sec, g in m/sec^2, H in m, and A in m^2. For Q in ft^3/sec, express V in ft/sec, g in ft/sec^2, H in ft, and A in ft^2.

C values are the same for SI or inch-pound units.

For intermittently moved systems, Equation 2 can be used to determine the gross average application rate. The application rate and time duration of each set will depend on evapo-transpiration, crop-rooting depth, soil water-holding capacity, and application efficiency.

$$R = \frac{60,000\ Q}{SL} \qquad (2)$$

where R is average application rate, mm/hr, Q is sprinkler discharge, m^3/min, S is between sprinkler spacing, m, and L is lateral spacing, m.

For traveling systems, Equations 3 through 5 can be used for system design. Manufacturer recommendations for wind speed-lane spacing relationships, if available, can be used instead of Equation 5.

$$D = \frac{1000\ Q}{ST} \qquad (3)$$

where D is average gross depth applied, mm, Q is sprinkler discharge m³/min, S = lane spacing, m, and T = travel speed, m/min.

$$A = 0.006 \, TS \qquad (4)$$

where A is area irrigated, ha/hr, T is travel speed, m/min, and S is lane spacing, m.

Equation 5 can be used to estimate travel-lane-spacing reduction, percent from no-wind spacing, required to obtain a Cu of 85% most of the time for wind speed that varies from zero to the maximum expected during the irrigation season.

$$\% \text{ red.} = 84 - 100/(\text{max wind speed, m/sec})^{1/2} \qquad (5)$$

REFERENCES

1. **Pichon, J. D.,** The Technology and Management of Traveling Sprinklers, in Proc. National Irrigation Symp., Section EE1, sponsored by American Society of Agricultural Engineers and Department of Agricultural Engineering, University of Nebraska, Lincoln, Nov. 10-13, 1970.
2. **Hanson, R. E.,** Water Distribution with Traveling Gun Sprinklers, Reprint No. 1321, presented at ASCE National Water Resources Eng. Meeting, Phoenix, Ariz., Jan. 11-15, 1971.
3. Nelson Irrigation Corp., Nelson Big Guns for Solid Set, Permanent Set, and Portable Sprinkler Systems, Nelson Irrigation Corp., Walla Walla, Washington 99362, undated, 6.
4. Nelson Irrigation Corp., Traveling Sprinkler System Planning Guide, No. TS-270, L. R. Nelson Mfg. Co., Peoria, Ill., 61614.
5. **Shull, H. and Dylla, A. S.,** Wind effects on water application patterns from a large, single nozzle sprinkler, *Trans. ASAE,* 19, 501, 1976.
6. **Shull, H. and Dylla, A. S.,** Operating Large Traveling "Gun" Sprinklers in Winds, paper No. 76-2014, presented at 1976 Annual Meeting of the American Society of Agricultural Engineers.
7. **Wiersma, J. L.,** We are testing the performance of large sprinklers in South Dakota, S.D. Farm Home Res., IX, 30, 1958.
8. **Allred, E. R. and Machmeier, R. E.,** Effect of wind resistance on rotational speed of boom sprinklers, *Trans. ASAE,* 5, 218, 1962.
9. **Machmeier, R. and Allred, E. R.,** Operating performance of a boom sprinkler, a field study, *Trans. ASAE,* 5, 220, 1962.

DRIP IRRIGATION SYSTEMS

I. P. Wu, C. A. Saruwatari, and H. M. Gitlin

INTRODUCTION

Drip (trickle) irrigation is a method of watering plants frequently and with a volume of water approaching the consumptive use of the plants. It differs from surface and sprinkler irrigation systems due to its slow water application rate. A drip irrigation system applies water in discrete or selected areas of a field by a system of small-diameter plastic lateral lines with holes called orifices or inserted devices called emitters at a selected spacing to deliver water to the soil surface near the base of the plants at a considerably lower water pressure than a sprinkler system.

The object of a drip irrigation system is to moisten the soil in the area of the crop roots only or to control the volume of soil occupied by the majority of the crop roots. Water movement in the soil is intended to be primarily through capillary action with a minimum of gravity movement. Drip irrigation is used to maintain a relatively constant soil moisture regime or at least one that does not vary to the extremes frequently encountered in surface and sprinkler irrigation.

A drip irrigation system consists of laterals, submains, and main lines. The laterals can be small plastic tubes with emitters or simply a small thin-walled plastic tube with orifices. The laterals are designed for distributing water into the field with an acceptable degree of uniformity. The submain acts as a control system, which can adjust the water pressure in order to deliver the required amount of flow into each lateral; it is also used to control irrigation time for individual fields. The main line serves as a conveyance system for delivering the total amount of water for the drip irrigation system.

The most important feature of drip irrigation is the slow water application rate at discrete or selected areas served by the operating system. Thus relatively small diameter "pipes" can be used to irrigate considerable acreages at one time. These "pipes" are the lateral lines, plastic (usually polyethylene or polyvinyl-chloride (PVC)) tubes usually 1/2- or 3/4-in. (nominal) diameter. Water is emitted from these laterals either through uniformly spaced orifices (0.012- to 0.025-in. diameter) burned or punched into the lateral or through discrete and separate emitters inserted into the lateral line at the desired locations. The flow rate from an orifice or emitter is generally between 0.3 to 1.0 gal/hr. To insure that the flow rate is kept low, the operating pressure for a drip irrigation system is also kept low (usually 8 to 12 lb/in.²)

In general, most laterals have either the simple orifice emitter or some form of a long-path emitter. The simple orifice emitter can be found on two general types of laterals — the single wall and the biwall lateral. In a single wall lateral, uniformly spaced burned or punched orifices in a thin-walled plastic tube allow water to be emitted at the slow application rate characteristic of drip irrigation systems. It operates at a water pressure of 8 to 12 lb/in.² The biwall lateral consists of a main tube that has a much smaller tube integral with it, somewhat like a figure eight. For each orifice leading from the main tube (primary chamber) to the smaller tube (secondary chamber) to supply it with water, there are four or more orifices in the smaller tube (secondary chamber) that allow water to be applied to the plants. The primary chamber operates at a water pressure of 12 to 15 lb/in.² while the secondary chamber emits water at a water pressure of less than 1 lb/in.² Therefore, it is possible to have larger orifices than the single wall lateral and still keep emission at a low level.

The long-path emitter derives its name from the long-path or maze system used to reduce the water pressure and emission rate. The simplest long-path emitter is the microtube, usually a polyethylene tube 0.036- to 0.1875-in. inside diameter. A selected (designed) length is thrust into a polyethylene or PVC lateral through a punched or drilled hole about 0.010 in. smaller than the microtube's outside diameter at a specified spacing. The water pressure drop due to friction caused by the long-path emitter provides the low pressure for the slow application rate in a drip irrigation system; it is also developed by inducing a swirl in some types of emitters.

The emitters, laterals, submains, and main lines are considered as the principal parts of a drip irrigation system. There are supporting parts such as filters, flushing units, pressure regulators, pressure gages, fittings, valves, and fertilizer injectors that are used in a drip irrigation system. Filters are very important in drip irrigation and are usually required in all cases. Since the orifices and emitters are very small, sediment, algae and other microorganisms, and chemical precipitates can cause plugging of the orifices. Therefore, a first consideration of a drip irrigation system is water quality or the cost of obtaining water of an acceptable quality. Flushing units (flush valves) at the end of the lateral line are used to flush out retained sediments and sludge. In some cases, over-pressuring the laterals is done to "blow out" the plugged orifices to clean them. Other systems can have a chlorine or other chemical injection system to clean the filters and the laterals. The addition of nutrients to the soil using a fertilizer injector in a drip irrigation system is known as fertigation.

BASIC HYDRAULICS

A drip irrigation system is made by a combination of different sizes of plastic pipes, which are usually considered as smooth pipes. One empirical equation that is frequently used is the Williams and Hazen formula for smooth pipe (using $C = 150$)

$$\Delta H = 9.76 \times 10^{-4} \frac{Q^{1.852}}{D^{4.871}} \Delta L \tag{1}$$

in which ΔH = energy drop by friction, in feet; Q = total discharge in the pipe, in gallons per minute (gal/min); D = inside diameter of the pipe, in inches; and ΔL = length of a pipe section, in feet. Equation 1 is used to determine the energy drop for a main-line section.

The flow condition in a lateral or a submain is steady and spatially varied with lateral outflows. Since the total discharge in the line decreases with respect to the length, the total energy drop will be less than given in Equation 1. The total energy drop, ΔH, for lateral or submain can be expressed as

$$\Delta H = 3.42 \times 10^{-4} \frac{Q^{1.852}}{D^{4.871}} L \tag{2}$$

in which ΔH = total energy drop by friction at the end of the lateral (or submain), in feet; Q = total discharge at the inlet of the lateral (or submain), in gallons per minute (gal/min); D = inside diameter of the lateral (or submain), in inches; and L = the total length of the lateral (or submain), in feet.

Since the total discharge in the lateral (or submain) decreases with respect to the length, the energy gradient line will not be a straight line but a curve. The shape of the energy gradient line can be expressed by the dimensionless energy gradient line

$$R_i = 1 - (1 - i)^{2.852} \tag{3}$$

in which $R_i = \Delta H_i/\Delta H$ is called the energy drop ratio; ΔH is the total energy drop determined by Equation 2, in feet; ΔH_i is the total energy drop, in feet, at a length ratio i ($i = \ell/L$); L = total length, in feet; and ℓ = a given length measured from the head end of the line, in feet. Equation 3 can be used to determine the energy drop pattern along a lateral (or submain) when the total energy drop, ΔH, is determined and known.

If a drip irrigation line, lateral or submain, is laid on level ground, the pressure distribution along the line will be indicated by the energy gradient line. If a drip irrigation line is laid on up or down slopes, the pressure distribution will be affected by the slopes. The pressure distribution (the change of pressure with respect to length) can be determined as a linear combination of energy slope and line slope. This can be expressed as

$$H_i = H - R_i\Delta H \pm i\Delta H' \tag{4}$$

where H_i = water pressure head at section i, in feet; H = operating (input) water pressure, in feet; R_i = energy drop ratio; ΔH = total energy drop, in feet; i = the length ratio; $\Delta H'$ = total energy gain (+) or loss (−) due to slope. Equation 4 is applicable only for a uniform slope situation, plus sign means downslope and minus means upslope. For a non-uniform slope situation, the references should be consulted.

Hydraulically, the pressure variation along a lateral line will cause an emitter flow variation along the lateral, and a pressure variation along a submain will cause a lateral flow variation (into each lateral line) along a submain. For the most common orifice emitters, and assuming turbulent flow in the laterals, the emitter flow (or lateral flow for submain) and the pressure head can be expressed by a simple function

$$q = C \sqrt{H} \tag{5}$$

in which q = emitter flow (or flow into lateral), H is the pressure head, and C is a constant. The pressure variation and emitter flow (or lateral flow) variation are related and can be expressed as

$$q_{var} = 1 - (1 - H_{var})^{0.5} \tag{6}$$

The emitter flow variation q_{var} is defined as

$$q_{var} = \frac{q_{max} - q_{min}}{q_{max}} \tag{7}$$

in which q_{max} is maximum emitter flow and q_{min} is the minimum emitter flow along the lateral line (or maximum and minimum lateral flow for submain). The pressure variation is defined as

$$H_{var} = \frac{H_{max} - H_{min}}{H_{max}} \tag{8}$$

in which H_{max} and H_{min} are maximum and minimum pressure head, respectively, along the lateral (or submain).

The output of drip emitters is affected primarily by the pressure in the lateral. Because the energy gradient line is already a curve, its combination with slope will magnify the difference in output between emitters. Since a plant may depend on one or more emitters for its water supply the variability in emitter discharge is very important in drip irrigation. A drip irrigation system must be designed to minimize the emitter flow variation or keep it

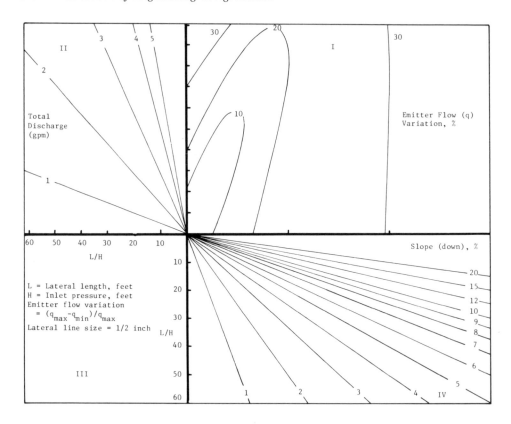

FIGURE 1.   Design chart for a 1/2-in. lateral line (down slope).

at an acceptable level. In drip irrigation design, the design criterion based on emitter flow variation is an emitter flow variation less than 20% (about 40% pressure variation) for a lateral line and a lateral flow variation less than 5% (about 10% pressure variation) for a submain design.

DESIGN PROCEDURES

Design charts were developed for lateral, submain, and main line design. The design charts and their design procedures are given as follows.

Lateral Line Design Charts for Uniform Slopes

The lateral line design charts were developed for a commonly used size, 1/2-in. (nominal) as shown in Figures 1 and 2.

Design Example 1 — The operating pressure of a lateral line is 6.5 lb/in.[2] (or 15 ft), the length of lateral line is 300 ft, the total discharge is 2 gal/min, the lateral line slope is 2% (downslope) and the lateral line size is 1/2 in. (I.D. 0.625 in.). Check the acceptability of the design.

Design Procedure:

The solution can be read from Figure 1 by the following procedure:

Step 1. Establish the lateral length (L) and operating pressure head (H) ratio L/H and total discharge (Q) in gallons per minute (gal/min).

$$L/H = 20 \text{ and } Q = 2 \text{ gal/min}$$

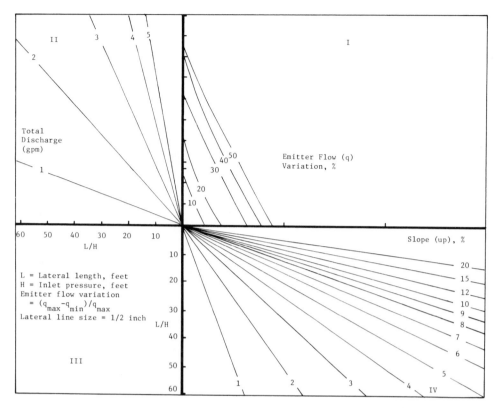

FIGURE 2.   Design chart for a 1/2-in. lateral line (up slope).

Step 2.   Move vertically from L/H (in Quadrant III) to the given total discharge gal/min line in Quadrant II; then establish a horizontal line into Quadrant I.

Step 3.   Move horizontally from L/H (Quadrant III) to the percent slope line in Quadrant IV; then establish a vertical line into Quadrant I.

$$\text{slope} = 2\% \text{ down}$$

Step 4.   The intersection point of these two lines in Quadrant I determines the acceptability of the design. The intersection point shows an emitter flow variation = 13%. The design is accepted. Note: Desirable (emitter flow variation less than 10% or pressure variation less than 20%). Acceptable (emitter flow variation about 10 to 20%, or pressure variation from 20 to 40%). Not recommended (emitter flow variation larger than 20%, or pressure variation greater than 40%).

A General Design Chart for Lateral and Submain on Uniform Slopes

A general design chart was developed by using a dimensionless term, total friction drop and length ratio, $\Delta H/L$, in Quadrant II of the lateral line design chart (Figures 1 and 2). These revised design charts as shown in Figures 3 and 4 are dimensionless and can be used for all pipe sizes. They are applicable for both lateral and submain designs. A nomograph for determining $\Delta H/L$ from total discharge and pipe size is given in Figure 5. This set of design charts can be used to check the acceptability of a design if the lateral (or submain) size is given, or to select a proper size of lateral (or submain) to meet the design criterion.

To check acceptability of a design — Lateral (or submain) size is given.

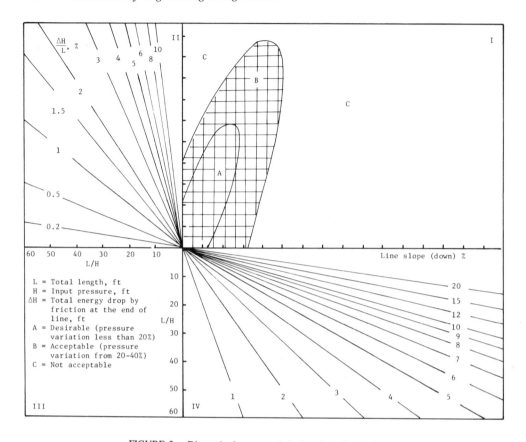

FIGURE 3. Dimensionless general design chart (down slope).

Step 1. Establish L/H and total discharge (gal/min).

Step 2. From the nomograph (Figure 5), use the total discharge and lateral line (or submain) size to determine ΔH/L.

Step 3. Move vertically from L/H (Quadrant III) to the determined ΔH/L in Quadrant II (Figure 3 or 4); then establish a horizontal line into Quadrant I.

Step 4. Move horizontally from L/H to the percent slope line in Quadrant IV; then establish a vertical line into Quadrant I.

Step 5. The intersection point of these two lines in Quadrant I determines the acceptability of the design:

Zone A = Desirable (emitter flow variation less than 10%)

Zone B = Acceptable (emitter flow variation from 10 to 20%)

Zone C = Not recommended (emitter flow variation greater than 20%).

Example 1 can be verified by using this procedure.

To Select Proper Lateral (or Submain) Size

A general design chart was developed by using a dimensionless term, total friction drop and length ratio, ΔH/L, in Quadrant II of the lateral line design chart (Figures 1 and 2). These revised design charts as shown in Figures 3 and 4 are dimensionless and can be used for all pipe sizes. They are applicable for both lateral and submain designs. A nomograph for determining ΔH/L from total discharge and pipe size is given in Figure 5. This set of design charts can be used to check the acceptability of a design if the lateral (or submain) size is given, or to select a proper size of lateral (or submain) to meet the design criterion.

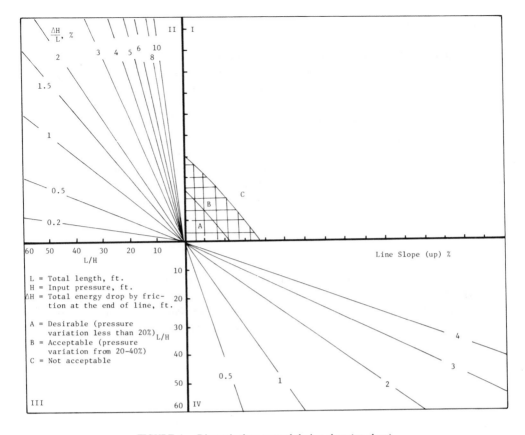

FIGURE 4. Dimensionless general design chart (up slope).

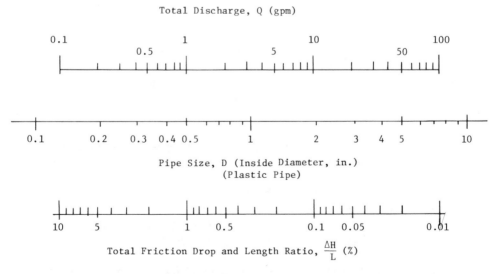

FIGURE 5. Nomograph for drip irrigation laterals and submain design in British units.

To check acceptability of a design — Lateral (or submain) size is given.

Step 1. Establish L/H and total discharge (gal/min).
Step 2. From the nomograph (Figure 5), use the total discharge and lateral line (or submain) size to determine ΔH/L.
Step 3. Move vertically from L/H (Quadrant III) to the determined ΔH/L in Quadrant II (Figure 3 or 4); then establish a horizontal line into Quadrant I.
Step 4. Move horizontally from L/H to the percent slope line in Quadrant IV; then establish a vertical line into Quadrant I.
Step 5. The intersection point of these two lines in Quadrant I determines the acceptability of the design: Zone A = Desirable (emitter flow variation less than 10%) Zone B = Acceptable (emitter flow variation from 10 to 20%) Zone C = Not recommended (emitter flow variation greater than 20%).

Example 1 can be verified by using this procedure.

To Select Proper Lateral (or Submain) Size
Design Example 2 — A lateral line length in a vegetable field is 150 ft and the slope of lateral line is 1% downslope. Emitters spaced 1 ft apart are installed in the lateral line. The emitter flow is 1 gal/hr at an operating pressure of 15 lb/in.^{2} Design the lateral line size.
    Given Information:
        Lateral length L = 150 ft
        Operating pressure H = 15 lb/in.^{2} = 34 ft
        Number of emitters = 150
        Total discharge Q = 150 gal/hr = 2.5 gal/min
    Design Procedure:

Step 1. Establish L/H and total discharge (gal/min).

$$L/H = 150/34 = 4.4 \text{ and } Q = 2.5 \text{ gal/min}$$

Step 2. Move horizontally from L/H to percent of slope line (down or up) in Quadrant IV; from that point establish a vertical line into Quadrant I.

$$\text{slope} = 1\% \text{ down}$$

Step 3. Establish a point along this line in Quadrant I at the upper boundary of the "desirable" region or "acceptable" region depending on the design criterion; from that point establish a horizontal line into Quadrant II. The desirable region (Zone A) should be used unless otherwise specified.
Step 4. Establish a vertical line in Quadrant II from the L/H value so that it intersects the horizontal line of Step 3 above, at a point.
Step 5. Determine the ΔH/L value in Quadrant II at this point. From Figure 3 and using the desirable zone (Zone A) as the design criterion, a ΔH/L value is determined as 6.
Step 6. From the nomograph (Figure 5) using total discharge and ΔH/L value, establish the minimum lateral size according to the selected design criterion. The minimum lateral size is determined as 0.5 in. (I.D.).

Design Example 3 — A subplot of a sugar cane field is 1 acre in size (rectangular). The submain has a length of 100 ft and is on a zero slope. The total discharge required for the

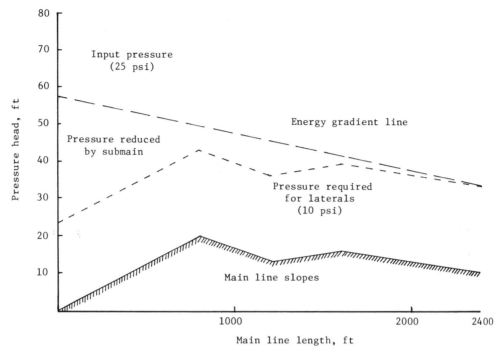

FIGURE 6.   Main line layout and energy gradient line.

submain to deliver is 15 gal/min and the operating pressure is 15 lb/in.[2] Design the submain size.

Design Procedure:

1. Total discharge: $Q = 15$ gpm
2. Submain length and pressure head ratio: L/H = 100/34 = 3
3. Submain slope: 0
4. From Figure 3 ΔH/L is found to be 10 and from Figure 5 the pipe size is designed as 1 in.

Main Line Design

A main line design is not a problem if it is a simple drip irrigation system in which a main line delivers water for only one or two subfields. This main line is designed based on the total discharge required in the main, the length and the allowable energy drop by using Equation 1.

When a main line delivers water into many submains (or many subfields), the total discharge in the main-line section decreases with respect to the length of main line. The size of a main line for different sections will depend on the shape of the energy gradient line above the main line. A straight energy gradient concept was developed for main line design that simplifies the design procedure greatly and can also be used for designing alternative field layouts of main lines.

Design Example 4 — A drip irrigation system is designed for a 50-acre papaya field. The area is rectangular in shape and divided into 1 acre subplots that are irrigated by a submain from the main line. The main line is laid in the center of the field with 25 acres on both sides of the main line. Each subplot is about 435 ft long and 100 ft wide; each main line section is 100 ft long. The design capacity is 30 gal/min/acre. There is a total of 24 sections and at the end of each section there will be an outlet to supply 60 gal/min for irrigating both sides. If the main line slopes are plotted as shown in Figure 6 and the required water pressure for lateral line is 10 lb/in.[2], design the main line.

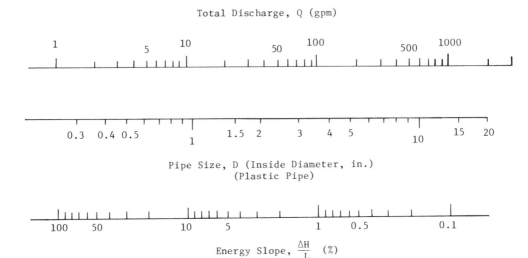

FIGURE 7.   Nomograph for drip irrigation, main line design in British units. (Calculated by Williams and Hazen Formula, C = 150.)

Design Procedure:

Step 1.   Plot main line profiles (topographical) as shown in Figure 6. Plot the required pressure, 10 lb/in.^{2}, along the main line, as 23 ft. above the ground profile.

Step 2.   Draw a straight energy gradient line from the available operating pressure to the required profile (Figure 6) so that everywhere along the main line the energy gradient line is above the required pressure profile.

Step. 3.   Determine the energy slope which is the slope of the straight energy gradient, $\Delta H/L$. From Figure 6, the energy slope is determined as 1%.

Step 4.   Determine required discharge in each main line section.

Step 5.   Design main line size by using nomograph, Figure 7, based on the energy slope (determined in Step 3) and total discharge (determined in Step 4) for each main line section. The main line sizes can be determined from the design chart, Figure 7, using 1% energy gradient line. The results are as follows:

Main line section	Discharge (gal/min)	Design diameter from Figure 7 (in.)
0[a]	1500	
1	1440	10
2	1380	10
3	1320	10
4	1260	8
5	1200	8
6	1140	8
7	1080	8
8	1020	8
9	960	8
10	900	8
11	840	8

12	780	8
13	720	8
14	660	8
15	600	6
16	540	6
17	480	6
18	420	6
19	360	6
20	300	5
21	240	5
22	180	4
23	120	4
24	60	3

[a]   There is an outlet at the entrance of section 1 for irrigating the subfields on both sides of section 1.

Special Design Problems

The design procedures presented in the previous sections are for uniformly sloped rectangular fields only. Design for nonuniformly sloped fields, nonrectangular fields, multiple pipe sizing, and other more complex situations are beyond the scope of this introduction to drip irrigation. The references should be consulted (especially References 7 and 9) for more details.

REFERENCES

1. **Howell, T. A. and Hiler, E. A.,** Designating trickle irrigation laterals for uniformity, Proc. Paper 10983, *J. Irrig. Drain. Div. ASCE,* 100(IR4), 443, 1974.
2. **Howell, T. A. and Hiler, E. A.,** Trickle irrigation lateral design, *Trans. ASAE,* 17(5), 902, 1974.
3. **Myers, L. E. and Bucks, D. A.,** Uniform irrigation with low pressure trickle system, Proc. Paper 9175, *J. Irrig. Drain. Div. ASCE,* 98(IR3), 341, 1972.
4. **Williams, G. S. and Hazen, A.,** *Hydraulic Tables,* 3rd ed., John Wiley & Sons, New York, 1960.
5. **Wu, I. P. and Gitlin, H. M.,** Design of Drip Irrigation Lines, Tech. Bull. No. 96, Hawaii Agric. Exp. Stn., University of Hawaii, Honolulu, June 1974.
6. **Wu, I. P. and Gitlin, H. M.,** Drip Irrigation System Design in Metric Units, Misc. Publ. No. 144, Coop. Ext. Serv., University of Hawaii, Oahu, March 1977.
7. **Wu, I. P. and Gitlin, H. M.,** Drip Irrigation System Design in British Units, Misc. Publ. No. 156, Coop. Ext. Serv., University of Hawaii, Honolulu, October 1978.
8. **Wu, I. P. and Gitlin, H. M.,** Revised Design Charts for Microtube Emitters in Drip Irrigation, The Engineer's Notebook No. 40, Coop. Ext. Serv., University of Hawaii, Honolulu, September 1978.
9. **Wu, I. P., Howell, T. A., and Hiler, E. A.,** Hydraulic Design of Drip Irrigation Systems, Tech. Bull. No. 105, Hawaii Agricultural Experiment Station, University of Hawaii, Honolulu, December 1979.

IRRIGATION PRACTICES AND OPERATIONS

R. R. Bruce and Jerry L. Chesness

SOIL/PLANT/WATER RELATIONSHIPS

In concept, irrigation should add prescribed amounts of water to a specified soil-water control zone at less than a specified soil-water control level as measured at a specified depth. It may seem incongruous that these are apparently soil-related quantities after emphasizing the plant as the focal point. However, in reality, plant characteristics and requirements strongly influence the value assigned to each. Since these two concepts are the keys to setting irrigation practice, a thorough understanding of them is necessary.

The soil-water control zone is the soil depth to which prescribed soil-water levels are to be maintained. Both soil and crop characteristics are used in setting this depth. Begin by establishing the rooting extent of the particular crop in the particular soil. When the crop has a naturally shallow or limited root system or when roots are restricted by soil characteristics, then this depth becomes the soil-water control zone. It likely represents 95 to 100% of the root system. In other instances, the crop may naturally have an extensive root system, and it may be distributed to great depths since the roots are not greatly restricted by soil conditions. The depth of the control zone in these situations is determined by the retention and flow characteristics of the soil that affect the leaching of mobile plant nutrients. Tables 1 and 2 give maximum depths available for water regulation in typical soil-texture groups in the Southern Coastal Plain and Southern Piedmont. Table 3 gives data for setting the soil-water control zone for selected crops.

The soil-water control level is the prescribed soil-water suction at a point, points, or region in the soil-water control zone at which irrigation is initiated. It is the highest soil-water suction allowable at a specified point, points, or region in the soil-water control zone (Table 3). Assignment of the soil-water control level depends upon crop species, depth and characteristics of soil-water control zone, fertilization practice, and method of water application. In this climatic region, the quantity of mobile nutrients leached out of the root zone can be reduced by scheduling application of fertilizer in relation to crop need and applying only enough irrigation water to recharge the soil-water control zone. Therefore, soil-water control level and the points of measurement must accurately reflect the available water stored in the water-control zone to avoid excessive water application.

The depth of soil-water control zone and the soil-water control level are interdependent — one must be adjusted in relation to the other to achieve the desired pattern of soil-water suction from irrigation practice. The relationship is illustrated by data from a potato irrigation experiment (Figure 1). In this case, it seems that 12 in. (30 cm) can be considered the soil-water control zone, and the soil-water control level is 0.5 bar soil-water suction at 6 in. (15 cm). Each time the 6-in. depth approached 0.5 bar, enough water was applied to wet the 6-in. depth to near 0.05 bar and the 12-in. depth to near 0.1 bar. Since the 12-in. depth was never wetted to the same level as the 6-in. depth, very little drainage occurred below 12 in. In this case, suction very seldom exceeded 0.5 bar in the 6-in. depth and did not exceed about 0.3 bar in the 12-in. depth. Nearly the same soil-water suction ranges could have been realized by irrigating when the 12-in. depth reached about 0.25 bar, although it is clearly a less sensitive indicator point than at 6 in., and it would be more difficult to maintain the 6-in. depth at 0.5 bar or less. This is due to the greater rates of water uptake at the 6-in. depth because of root concentration. The mean of the readings at the two depths might have been used to signal the start of irrigation. However, a value of the mean must be selected to assure that excessively high soil-water suctions do not occur in the zone of major root concentration and rapid water extraction.

Table 1
CHARACTERISTICS OF SOIL-TEXTURE GROUPINGS FOR SOUTHERN PIEDMONT SOILS

Soil-texture grouping	Layer	Soil texture	Maximum depth available for water regulation (in.)[a]	Water retention, in./in., at suction of —						Maximum water application rate (in./hr)[b]
				0.03 bar	0.06 bar	0.25 bar	0.50 bar	0.75 bar	1.0 bar	
A	Surface	Loamy sand or coarse sandy loam	7	0.22	0.17	0.125	0.11	0.105	0.10	0.5
	Subsoil	Sandy clay loam, clay loam, or clay	11	0.36	0.34	0.32	0.30	—	0.29	
B	Surface	Sandy loam	6	0.28	0.23	0.17	0.165	0.155	0.15	0.5
	Subsoil	Sandy clay loam, clay loam, or clay	12	0.36	0.34	0.32	0.30	—	0.29	
C	Surface	Loam to clay loam	6	0.35	0.34	0.32	0.31	0.30	0.295	0.3
	Subsoil	Sandy clay loam, clay loam, or clay	12	0.36	0.34	0.32	0.30	—	0.29	

[a] To convert inches to centimeters, multiply by 2.54.
[b] To convert inches per hour to centimeters per hour, multiply by 2.54.

Table 2
CHARACTERISTICS OF SOIL-TEXTURE GROUPINGS FOR SOUTHERN COASTAL PLAIN SOILS

Soil-texture grouping	Layer[a]	Soil texture	Water retention, in./in., at suction of —						Maximum water application rate (in./hr)[b]
			0.025 bar	0.05 bar	0.10 bar	0.25 bar	0.50 bar	1.0 bar	
A	Surface	Sand and loamy sand	0.29	0.20	0.13	0.10	0.08	0.07	0.7
	Subsoil	Sand and loamy sand	0.29	0.20	0.13	0.10	0.08	0.07	
B	Surface	Sand and loamy sand	0.29	0.20	0.13	0.10	0.08	0.07	0.7
	Subsoil	Sandy loam and fine sandy loam	0.31	0.26	0.20	0.17	0.15	0.13	
C	Surface	Sand and loamy sand	0.29	0.20	0.13	0.10	0.08	0.07	0.5
	Subsoil	Sandy clay loam and sandy clay	—	0.30	0.27	0.25	0.23	0.22	
D	Surface	Loamy fine sand	0.29	0.25	0.18	0.13	0.11	0.09	0.5
	Subsoil	Sandy clay loam and sandy clay	—	0.30	0.27	0.25	0.23	0.22	
E	Surface	Loamy fine sand	0.29	0.25	0.18	0.13	0.11	0.09	0.5
	Subsoil	Sandy loam and fine sandy loam	0.31	0.26	0.20	0.17	0.15	0.13	
F	Surface	Sandy loam and fine sandy loam	0.31	0.26	0.20	0.17	0.15	0.13	0.5
	Subsoil	Sandy clay loam and sandy clay	—	0.30	0.27	0.25	0.23	0.22	

[a]   Maximum layer thickness available for water regulation is 10 in. (25 cm).
[b]   To convert inches per hour to centimeters per hour, multiply by 2.54.

Table 3
SOIL-WATER CONTROL ZONE AND MAXIMUM SOIL-WATER SUCTION FOR SELECTED CROPS

Crop	Depth of soil-water control zone		Maximum soil-water suction in control zone[a] (bar)
	Inches	Centimeters	
Corn	12	30	0.6
Cotton	12	30	0.6
Cucumbers	9	23	0.5
Peaches	16	41	0.6
Peanuts	18	46	0.6
Pecans	18	46	0.6
Snap beans	9	23	0.4
Southern peas	12	30	0.6
Soybeans	18	46	0.6

[a]   This is the maximum mean soil-water suction that should be allowed by the adopted irrigation procedures. In practice, the soil-water suction at which irrigation should begin is less than these values and is affected by the particular soil and water regulation system.

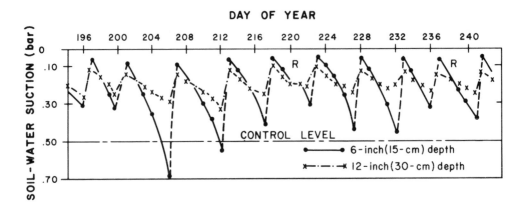

FIGURE 1.   Tensiometer record of soil-water suction in potatoes. Irrigations were scheduled when water suction reached 0.5 bar at the 6-in. (15-cm) depth. R indicates less than 0.18 in. (5 mm) of rainfall. (From Taylor, Sterling A. and Ashcroft, Gaylen L., *Physical Edaphology*, W. H. Freeman, San Francisco, 1972. With permission.)

In summary, the frequency of application depends upon the depth of the control zone, the specified soil-water suction limit, and the current water demands of the crop. In cases of shallow control zones maintained at low soil-water suctions on sandy soils, frequent small applications are required to adequately meet crop water requirements without risking excessive leaching of the more mobile nutrients. In other cases, water-control zones of greater depth and higher soil-water suctions may be satisfactory; therefore, larger water applications are made less frequently.

FERTILIZATION THROUGH IRRIGATION SYSTEMS

Satisfying a crop's nutrient requirements cannot be separated from satisfying its water requirements. Inadequate fertilization of irrigated crops in the Southeast has seriously limited yields and lowered return on investment. By regulating water supply to plants, all cultural practices become more critical, and more attention must be paid to adequate fertilization. This may even include attention to some essential nutrients for crop growth that were adequate in unirrigated culture. Therefore, the best information possible must be obtained in meeting plant nutrient needs.

In this region, the managing of mobile nutrients deserves particular attention. Since nutrients can easily be applied through the commonly used irrigation systems, crop requirements can be programed and met simultaneously with water requirements. By such a procedure, nutrient losses by leaching should be minimized and effectiveness of applied materials increased. Nitrogen fertilization by this means is partieularly attractive. The main advantages of applying nutrients with water are savings in labor and equipment, better timing and utilization of the nutrient application, ease of split and multiple applications, and larger yields of better quality crops.

With the great variety of fertilizer products available, there are several choices to be considered. Liquid fertilizers are convenient to handle with pumps and gravity flow. They may contain a single nutrient or combinations of nitrogen (N), phosphorus (P), and potassium (K). Liquid fertilizers commonly applied through irrigation systems are listed in Table 4. There is a wide variety of soluble dry fertilizers containing nitrogen, phosphorus, and potassium singly or in combination. Dry fertilizer can be dissolved by mixing with water in a separate, open tank and then pumped into the irrigation stream, or it can be placed in a pressurized container through which part of the sprinkler stream passes; the bypass water dissolves the fertilizer and carries it into the main stream. Typical dry fertilizers that may be applied through irrigation systems are listed in Table 5.

Table 4
LIQUID FERTILIZERS FOR APPLICATION THROUGH IRRIGATION SYSTEMS

Fertilizer	Formula	Total nitrogen (% N)	Available phosphoric acid (% P)	Water-soluble potash (% K)	Approximate pounds of product for 1 lb of nutrient[a]		
					N	P	K
Ammonium nitrate	NH_4NO_3	20	—	—	5	—	—
Ammonium phosphate	$NH_4H_2PO_4$	8	24	—	12	4	—
Potassium ammonium phosphate	KNH_4HPO_4	15	15	10	7	7	10
(N-P-K liquid mixes)		10	10	10	10	10	10
		15	8	4	7	12.5	25
Urea (low biuret)	$CO(NH_2)_2$	23	—	—	4.4	—	—
Urea-ammonium nitrate	35.4% $CO(NH_2)_2$ + 44.3% NH_4NO_3	32	—	—	3.1	—	—
Phosphoric acid (green)	H_3PO_4	—	52—54	—	—	1.9—1.8	—
Calcium ammonium nitrate	11.6% $Ca(NO_3)_2$ + 5.4% NH_4NO_3	17	—	—	6	—	—

[a] To convert pounds to kilograms, multiply by 0.454.

From Soil Improvement Committee, Western Fertilizer Handbook, California Fertilizer Association, Sacramento, 1961. With permission.

Table 5
DRY FERTILIZERS FOR APPLICATION THROUGH IRRIGATION SYSTEMS[a]

Fertilizer	Formula	Total nitrogen (% N)	Available phosphoric acid (% P_2O_5)	Water-soluble potash (% K_2O)	Total sulfur (% S)	Approx. pounds of product for 1 lb of nutrient[b]			
						N	P_2O_5	K_2O	S
Ammonium nitrate	NH_4NO_3	33.5	—	—	—	3	—	—	—
Calcium ammonium nitrate	$CaNH_4(NO_3)_3$	26	—	—	—	4	—	—	—
(Mono)ammonium phosphate	$NH_4H_2PO_4$	11	48	—	2.6	9	2	—	40
Ammonium phosphate sulfate	$NH_4H_2PO_4$ + $(NH_4)_2SO_4$	13	39	—	7	8	2.5	—	14
Ammonium phosphate sulfate	40% $NH_4H_2PO_4$ + 60% $(NH_4)_2SO_4$	16	20	—	15.4	6	6	—	7
Ammonium phosphate nitrate	$NH_4H_2PO_4$ + NH_4NO_3	24	20	—	—	4	6	—	—
Ammonium phosphate nitrate	$NH_4H_2PO_4$ + NH_4NO_3	27	14	—	—	4	7	—	—
Diammonium phosphate	$(NH_4)_2HPO_4$	21	53	—	—	5	2	—	—
Ammonium chloride	NH_4Cl	25	—	—	—	4	—	—	—
Ammonium sulfate	$(NH_4)_2SO_4$	20—21	—	—	24	5	—	—	4
Calcium nitrate	$Ca(NO_3)_2$	15.5	—	—	—	6	—	—	—
Sodium nitrate	$NaNO_3$	16	—	—	—	6	—	—	—
Potassium nitrate	KNO_3	13	—	44	—	8	—	2.3	—
Urea	$CO(NH_2)_2$	45—46	—	—	—	2.2	—	—	—
Double or treble superphosphate	$Ca(H_2PO_4)_2 \cdot H_2O$	—	42—46	—	10	—	2.3	—	10
Potassium chloride	KCl	—	—	60—62	—	—	—	1.7	—
Potassium sulfate	K_2SO_4	—	—	50—53	18	—	—	2	5.5
Sulfate potash magnesia	$K_2SO_4 \cdot 2MgSO_4$	—	—	26	15	—	—	4	7
Nitrate soda potash	$NaNO_3 \cdot KNO_3$	16	—	14	—	7	—	7	—

[a] Check with governmental chemical regulatory agencies for approval before use.
[b] To convert pounds to kilograms, multiply by 0.454.

From Soil Improvement Committee, Western Fertilizer Handbook, California Fertilizer Association, Sacramento, 1961. With permission.

The common methods of injecting fertilizers into an irrigation system are as follows:

Gravity mixing and injection on suction side of pump — A centrifugal pump obtaining water from a free water surface, such as a ditch or pond, develops a negative pressure in its suction pipe. This reduced pressure can be used to draw fertilizer solutions into the pump. A hose or pipe delivers the fertilizer from an open supply tank to the suction pipe. The rate of delivery is controlled by a valve. This connection must be tight to keep air from entering the pump.

Pressure pump injection — When turbine pumps with submerged impellers are used, the fertilizer (or other chemical) solution may be injected into the sprinkler line under pressure. A small separate rotary, gear, diaphragm, or piston pump can be used to force the chemical solution from the supply tank into the pressure line. This pump must be used to force the chemical solution from the supply tank into the pressure line. This pump must develop a greater pressure than that delivered by the irrigation pump. The internal parts of the pump should be noncorrosive. An external source of power for the injector pump must be provided.

Venturi principle of injection — A tapered construction known as a Venturi unit can be used to inject chemicals into a pipe line under pressure. A pressure drop accompanies the change in velocity of the water as it passes through the constriction. The difference in pressure between a connection above and one within the constriction is enough to cause a flow of water through a supply tank containing the chemicals. Since the water flowing through the tank is under pressure, a sealed airtight pressure supply tank constructed to withstand the maximum operating pressure is required.

Pitot tube injection — One open-ended pipe is faced into the stream of water, and another is faced opposite in the direction of the water flow. This arrangement establishes a flow of water between the pipe ends. The water is circulated through an airtight pressure supply tank containing fertilizer or other chemicals, similar to the Venturi unit.

MEASUREMENT OF SOIL-WATER SUCTION

Day-to-day scheduling of water application depends upon measurements of soil-water suction. This is an intrinsic property of the system by which plants are directly influenced. Plants are only influenced indirectly by soil-water content. Measurement of soil-water suction removes much of the subjectivity common in irrigation practice. When plants show visible signs of water deficit, plant growth has already been retarded. In some instances, it may be practical to automatically switch the irrigation system on and off from a soil-water suction measuring device. Otherwise, instrument readings must be recorded daily during periods of no rainfall. Daily readings allow one to project initiation of irrigation.

Soil-water suction is considered an equivalent pressure in the soil solution. Three instruments to measure soil-water suction are on the market: tensiometers, electrical resistance blocks, and matric potential sensors (soil-water suction is sometimes called matric potential).

Tensiometers — These instruments may be called irrometers, soil-moisture gages, or soil-moisture indicators. To be classified as tensiometers, they must directly measure equivalent hydrostatic pressure in unsaturated soil. They have two major components: a porous ceramic cup that allows water to pass, but when wet prevents air passage, and a gage that measures tension (vacuum or negative pressure) with respect to barometric pressure. The porous cup and gage are connected by tubing. Figure 2 shows a commercially available tensiometer. For pressure in the soil water to be adequately measured, the installation of the porous cup in the soil must insure contact between the water in the soil pores and the water in the pores of the ceramic cup. Water moves in and out of the tensiometer through the cup wall as hydraulic equilibrium is maintained between the water in the soil and in the tensiometer. The soil-water suction or tension can be read on the gage hydraulically connected

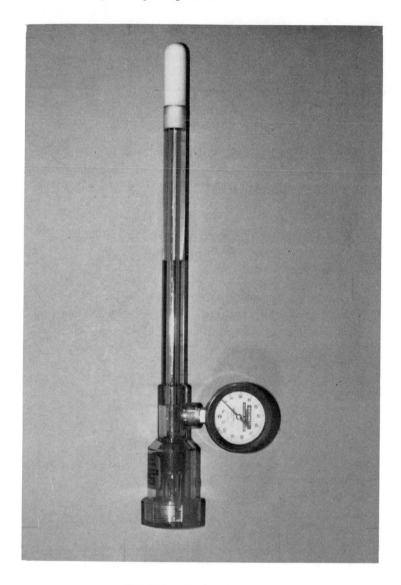

FIGURE 2. Commercial tensiometer.

to the cup. For a tensiometer to work, the pores of the ceramic cup must remain filled with water, and the tube between cup and gage remain free of air. The measurement range of the tensiometer is limited to less than 1 bar (in practice, less than about 0.85 bar), a range that is adequate for most irrigation practice. Gages on most commercially made tensiometers read from 0 to 100 cbars (1 bar = 100 cbars). If tensiometers are properly installed and used in their functional range, relatively little maintenance is required. However, each time they are read, the need for servicing should be determined. Tensiometers may be purchased with an electrical switching device that automatically turns an electrical system on and off at preset suctions. Alternatively, electrical switching tensiometers may be used to turn a system on, which may then be turned off by an interval timer.

Electrical resistance blocks — These devices consist of two electrodes embedded in a porous material such as gypsum, nylon, or fiberglass. When the porous material is placed in contact with the soil, water moves from soil to block or block to soil until the water in block and soil are in equilibrium. The electrical resistance measured between the electrodes

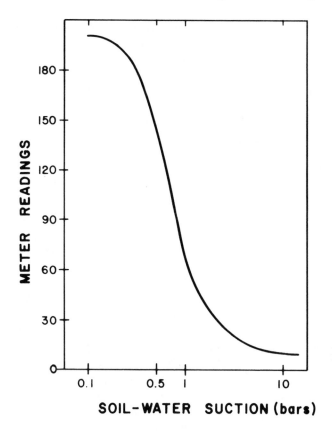

FIGURE 3.   Calibration of gypsum electrical resistance block
(Delmhorst).

is a function of the amount of water in the porous material. Since the nature of the pore system in the block material affects the quantity of water in the block at a given suction, the relation between electrical resistance and soil-water suction will be different for each type of porous material and porous block preparation. Assuming that blocks have been uniformly fabricated from a given material and electrodes embedded in a uniform geometry, the relation of block resistance to soil-water suction will be constant for a large variety of soils. The nature of this relation is shown in Figure 3. To eliminate the effect of the soil characteristics upon the measured electrical resistance, block design is very important. The use of cylindrical electrodes in cylindrical blocks and screen electrodes in rectangular blocks has eliminated this problem. Embedded electrodes in gypsum are most commonly used. Two or three manufacturers market resistance units and meters for quickly reading electrical resistance. To insure ready water transfer, contact between block and soil must be satisfactory. The soil must be carefully tamped back in the installation hole to a density slightly higher than natural density to prevent excess water at the surface from freely running down the hole to the blocks. A slight mounding at the surface is a further precaution. Installation instructions are available from manufacturers. Electrical resistance blocks will not measure soil-water suction reliably at less than 0.25 bar. However, they may be used in many situations. They can be read quickly and do not require much maintenance. Gypsum blocks will slowly dissolve and must be replaced every 1 or 2 years in the Southeast, but they are inexpensive.

Matric potential sensors — These devices also include a porous ceramic material in their construction. They must be installed at prescribed depths so that water readily moves between them and the soil to achieve equilibrium. Matric potential sensors measure the change in soil-water content of the porous block by determining the effect upon thermal conductivity. Devices in the block measure temperature change in the block resulting from an imposed heat pulse from a very small electrical heater. Currently available sensors measure most reliably in the range of 0.1 to 1 bar, but they may be constructed to measure at higher soil-water suctions. Maintenance is fairly low, and the ceramic lasts for several years. They cost more than electrical resistance blocks and tensiometers. The manufacturer offers a variety of meter types and controls for irrigation equipment.

SIZING AND OPERATING IRRIGATION SYSTEMS

Two types of water application systems, sprinkler and drip systems, are particularly well suited to the climate, soils, and topography of the Southeast. Before equipment is purchased, the system capacity required to meet the crop's needs must be determined. After equipment is installed, it must be operated appropriately to meet the crop water requirements.

Sprinkler Systems

Sprinkler systems, either center-pivot or cable-tow, apply water as droplets that range in diameter from 0.04 to 0.28 in. (0.10 to 0.71 cm), depending on nozzle design and operating pressure. Larger droplets are less subject to being haphazardly dispersed by wind and to evaporative losses, but they can injure small or delicate plants more and can accelerate surface crusting in certain soils. Once droplets reach the soil surface, they either infiltrate and move downward by gravity and capillary forces, or produce runoff. While some lateral movement occurs, most of the movement is downward. To avoid runoff, a sprinkler system should apply water uniformly over the surface of the entire field at a rate less than the infiltration rate. Uneven water distribution results in too little water being applied in some areas, reducing crop yield, and too much in other areas, resulting in deep percolation losses, runoff and perhaps erosion, higher energy costs, and reduced yields. The coefficient of uniformity (C_u) is the most commonly used measure of application uniformity. It compares application depths at selected locations to the mean value. Absolute uniformity is represented by a C_u of 100%. Systems with C_u values of 80% or higher are generally acceptable.

Application efficiency (E_a) is defined as the ratio of water retained in the water-regulation zone to the total water applied. It depends upon the losses from evaporation, runoff, and deep percolation. Since these losses generally represent water that is unavailable to the plants, the flow capacity of the system must be increased by the amount of these losses. Application efficiencies range from a low of 70% for high-volume traveling sprinklers under windy conditions to a high of 85% for low-pressure center-pivot systems under low-wind conditions.

A system's delivery capacity must be large enough to meet maximum daily potential evapotranspiration. Determining this capacity requires a knowledge of the area to be irrigated (A, in acres or hectares), the maximum evapotranspiration rate of the crop to be grown (ET, in inches or centimeters per day — see Table 6), the application efficiency of the system (E_a, as a decimal fraction), and the hours of operation per day (hr). The required capacity, Q_s, in gallons per minute or liters per second is

$$Q_s = (K \times ET \times A)/(E_a \times hr) \tag{1}$$

where K = 453 for U.S. customary units and 27.8 for metric units.

The soil-water recharge (SWR, in inches or centimeters) of the root zone depends upon the soil depth to which water is regulated (D, in inches or centimeters), and the water retention of the regulated zone at the low (WR_L, in inches per inch or centimeters per

Table 6
MAXIMUM EVAPOTRANSPIRATION RATES
FOR SELECTED CROPS

Crop	Evapotranspiration rate	
	In./day	Cm/day
Corn, cotton, peanuts, soybeans	0.3	0.76
Cucumbers, southern peas, snap beans	0.2—0.3[a]	0.51—0.76[a]
Peaches, pecans	0.2—0.3[b]	0.51—0.76[b]

[a]   Value depends on completeness of canopy.
[b]   Value depends on type and amount of ground cover and on tree canopy.

centimeter) and high (WR_h, in inches per inch or centimeters per centimeter) water-suction values selected. The equation expressing this relationship is

$$SWR = (WR_L - WR_h)(D) \tag{2}$$

If the soil in the zone of regulation is layered, Equation 2 should be applied to each layer and the results summed.

The maximum irrigation interval (I, in days) is calculated from the soil-water recharge (SWR, in inches or centimeters) and the peak evapotranspiration rate (ET, in inches or centimeters per day):

$$I = SWR/ET \tag{3}$$

This interval should not be used for scheduling irrigations, however, because of the variability in daily ET values and effective rainfall amounts experienced during the growing season. It serves only as a potential upper limit for estimating the maximum time between irrigations. Actual in-field scheduling should be based on measured soil-water suction within the water-regulation zone.

The power required to pump water for sprinkler systems is a function of the discharge capacity of the system and in turn that of the pump (Q_s, in gallons per minute or liters per second), the total pumping head, including suction lift and friction losses (H_t, in feet or meters), and the efficiency of the pump (E_p, as a decimal fraction). The brake horsepower requirement (BHP) relationship is:

$$BHP = (Q_s \times H_t)/(K \times E_p) \tag{4}$$

where K = 3960 for U.S. customary units and 76 for metric units.

Center-Pivot Systems

The center-pivot arrangement requires that each successive section of main line (moving from the center pivot to the outer end) must irrigate an increasingly larger area. To achieve this, the discharge capacity of the sprinklers (which are usually attached directly to the main line) increases with distance from the pivot point. Sprinkler wetting patterns are also overlapped to obtain the required application rates and uniformity. Generally, those systems with variable sprinkler spacings as opposed to fixed spacings provide a more uniform increase (center pivot to outer end) in application rate.

Table 7

MAXIMUM WATER APPLICATION RATES FOR
THREE DIFFERENT SPRINKLER ARRANGEMENTS
ALONG A CENTER-PIVOT LATERAL

Sprinkler arrangement	Maximum wetted diameter per sprinkler		Maximum application rate[a]	
	Feet	Meters	In./hr	Cm/hr
Variable-size sprinklers	175	53.3	1.3	3.3
All medium-size sprinklers	90	27.4	2.0	5.1
Spray-jet sprinklers	20	6.1	6.0	15.2

[a]   Measured at a distance of 1200 ft (366 m) from the pivot.

From Hagan, Robert M., Haise, Howard R., and Edminster, Talcott W., *Irrigation of Agricultural Lands,* American Society of Agronomy, Madison, Wis., 1967. With permission.

To achieve a uniform depth of water application from pivot to end point, the applicator must continuously increase the application rate by increasing the number of sprinklers, or increase the sprinkling time and rate by increasing the discharge and wetting diameters of the sprinklers. Most systems use a combination of these methods, but several manufacturers have begun offering a "low pressure" system with nonrotating sprinklers that utilizes only the first method. As the area irrigated per system increases, so also does the length of the main lines. This increases the possibility that peak application rates near the end of the main line will exceed the soil infiltration rate (see Table 7). If this occurs for a long enough time over a large enough area to produce surface runoff, water regulation and crop response can be adversely affected. If excessive application rates cannot be avoided through suitable sprinkler sizing and spacing, the size of the area irrigated per system should be reduced.

The total depth of water applied per irrigation by a center-pivot system is a function of the application rates along the main line and the travel speed. Since application rates are fixed by system design, the irrigator controls depth of watering by changing the system travel speed. All systems should have a travel speed that allows one full circle (or less if obstructed) in 24 hr and applies enough water to meet the peak ET demand of the crop. Applications less than this amount are often desirable for seed germination, seedling emergence, and chemical applications. In this situation, the system must have a full-circle travel time of less than 24 hr, for example, 12 hr if one half of the peak ET amount is to be applied. When the water-regulation plan for a particular crop and soil permit ET storage for more than 1 day in the water-regulation zone, the full-circle travel time can be adjusted to meet this longer time period if desired. Table 8 summarizes the range of operating and performance characteristics of currently available center-pivot systems. Because of the day-to-day variability in ET and rainfall experienced in this region, the irrigator must rely on measurements of soil-water suction in determining when and how much to irrigate. Tensiometers are well-suited for medium- to coarse-textured soils where suctions are to be maintained at less than 0.8 bar.

Two tensiometers should be placed at each of several stations: one in the center of the soil-water control zone, but not less than 6 in. (15 cm) deep to determine when to begin irrigation, and a second near the bottom of the soil-water control zone to determine whether or not the water applied per irrigation is recharging the whole zone.

Frequency of tensiometer readings will depend upon climatic conditions and crop growth

Table 8
PERFORMANCE CHARACTERISTICS
OF CENTER-PIVOT SYSTEMS

Characteristic	Operating range
End sprinkler operating pressure	20—80 lb/in.^{2} (138—552 kPa)
System size	20—400 acres (8—162 ha)
Application depth	0.1—5.0 inches (0.25—12.7 cm)
Travel time	10—96 hr per revolution
Type of drive	Electric or hydraulic

stage. We suggest reading them a minimum of three times a week at the beginning of the season. As the plants mature and water demands increase, daily readings may be necessary. The rate of change of soil-water suction between readings will dictate the frequency.

The number and location of tensiometer stations should be based on field size and soil variability, planned irrigation schedule, and accessibility. We suggest a minimum of six tensiometer stations for fields of 100 to 200 acres (40 to 81 ha). The irrigation schedule and full-circle travel time must also be considered when deciding where to locate the stations. If possible, there should be one or more stations within each area irrigated during a 24-hr period. Accordingly, if the full-circle travel time is 24 hr, stations can be located to best represent the soil variability in the field. If the full-circle travel time is 72 hr, placing two stations near each of three radii (120° apart) would be advisable. These stations might be located at points one third and three fourths radius lengths from the pivot. Allowances should be made, of course, for soil variability and station accessibility. Locating stations next to access roads or near support tower wheel tracks is convenient, although ease of access must not be the principal criteria in determining the number and location of the stations. The main objective is to obtain suction readings that best represent the average conditions throughout the entire field. Flagging and mapping the station locations is advisable.

EXAMPLE PROBLEMS AND SOLUTIONS

The intention is to illustrate the use of the principles and procedures that have been presented to determine appropriate water-regulation and associated cultural practices for satisfactory crop production in given farm situations. The specific land area and crop are first of all identified. This directs attention to knowledge of a given crop and its culture on a particular soil, in a particular climate, and where availability of quality water can be specified. Included as background information should be projected crop sequence and peculiar soil problems associated with past use, such as hardpans or erosion. Then crop culture and water regulation can be based on the given information and current knowledge of crop-soil-water relations.

Corn in Tift County, Ga.

A farmer in the vicinity of Tifton, Ga. plans to grow irrigated corn in a 160-acre (64.8-ha) field. Corn will be rotated with peanuts, and occasionally snap beans will be grown. The Tift County soil survey map[3] shows that over 80% of the field is a Tifton loamy sand (Plinthic Paleudult) having a 2 to 5% slope. The soil is generally well-drained, and where needed, surface drainage has been provided. This site has a potential erosion problem, and suitable control measures must be imposed. The surface soil is about 10 in. (25 cm) deep over about 7 in. (18 cm) of sandy loam to fine sandy loam. There are no pans or impedances to root exploration to the 17-in. (43-cm) depth. Long-term weather records indicate about a 50% chance of receiving 1 in. (2.5 cm) of rainfall per week from June 20 through July, with a much lower chance both before and after this period. Also there is a 60 to 70% chance that evaporation exceeds rainfall. Tift County is underlain by the Tampa, Suwannee, and Ocala Limestone Formations, which compose the principal artesian aquifer, and it is feasible to drill a well with the capacity and quality to supply a large sprinkler system.

In the process of procuring a satisfactory irrigation system, careful selection of corn variety, fertilization practices, water procedures, and other cultural aspects must not be deemphasized. In fact, a complete cultural plan becomes more important than in unirrigated culture. In view of the abundant water supply, field geometry, high infiltration capacity of the soil, and the farmer's plans to grow row crops with a minimum of labor, a center-pivot system is recommended.

The capacity of the well should be large enough to meet the peak water demands of the crop with the highest water use. In this example, peanuts, corn, and snap beans all have a peak daily water requirement (Table 6) of 0.3 in./day (0.76 cm/day). Assuming an application efficiency of 80% and a full-circle travel time of 24 hr, the required well and irrigation system capacity is

$$Q_s = (K \times ET \times A)/(E_a \times hr)$$

$$= (433 \times 0.3 \times 160)/(0.8 \times 24)$$

$$= 1132 \text{ gal/min (71 } \ell/\text{sec)}$$

Assuming a 100-ft (30.5-m) lift at the well (located at pivot point) with a pivot-point operating pressure of 60 lb/in.^{2} (414 kPa) for a total head of 236 ft (72 m) and a pump efficiency of 70%, the brake horsepower requirement is calculated as

$$BHP = (Q_s \times H_t)/(K \times E_p)$$

$$= (1130 \times 236)/(3960 \times 0.7)$$

$$= 96$$

This horsepower requirement could be met utilizing a direct-drive 100-hp electric motor or a 120-hp diesel engine.

Water Regulation

The depth of water regulation for growing corn on a loamy sand is assumed to be 12 in. (30.5 cm) (Table 3). Response to water regulation has been good when the water suction at the 6-in. (15-cm) depth has not been allowed to exceed 0.6 bar (Table 3). This suction level should not be considered as the starting point for irrigation because of the time required for the system to irrigate the entire field. Therefore, we will plan to irrigate each area in 24 hr or less when the mean soil-water suction at the 6-in. depth reaches 0.40 bar. This value will also be taken as the high suction value for calculating soil-water recharge. A value of 0.05 bar will be taken as the low value. From Table 2, soil-texture grouping B, we obtain water-retention values to calculate, using Equation 2, the soil-water recharge in the soil-water control zone between the suction limits of 0.05 and 0.40 bar:

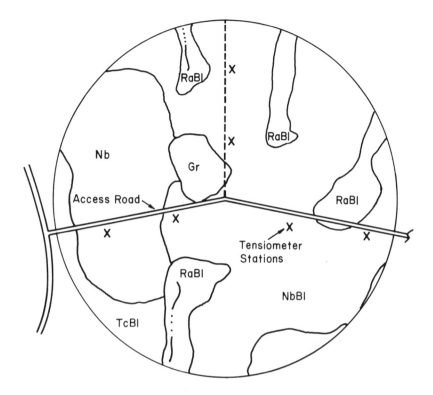

FIGURE 4. Proposed field site for 160-acre (65-ha) center-pivot irrigation system. Gr, Grady sand loam; Nb, Norfolk loamy sand, level thick surface phase; NbBl, Norfolk loamy sand, very gently sloping thick surface phase; TcBl, Tifton sandy loam, very gently sloping phase; RaBl, Rains loamy sand, very gently sloping thick surface phase.

$$SWR = (WR_L - WR_h)(D_1) + (WR_L - WR_h)(D_2)$$
$$= (0.20 - 0.09)(10) + (0.26 - 0.16)(2)$$
$$= 1.3 \text{ in. } (3.3 \text{ cm})$$

The time required for the corn plants (plus evaporation) to remove the stored water during periods of maximum evapotranspiration (0.3 in./day or 0.76 cm/day from Table 6) is calculated as

$$I = SWR/ET$$
$$= 1.3/0.3$$
$$= 4.3 \text{ or about 4 days}$$

This value can be used to estimate the maximum time between irrigations when peak evapotranspiration requirements must be met (actual scheduling should be based on in-field suction measurements) and to determine the maximum permissible full-circle travel time for the system.

Tensiometers should be installed to determine when to irrigate and how much water to apply. Consider a minimum of six tensiometer stations located on each of three radii approximately 120° apart. The field layout is shown in Figure 4, and note that there are two access roads to the pivot point. While these two roads do not exactly fit the 120° criteria, they will provide easy access to four of the six stations. The other two stations could be

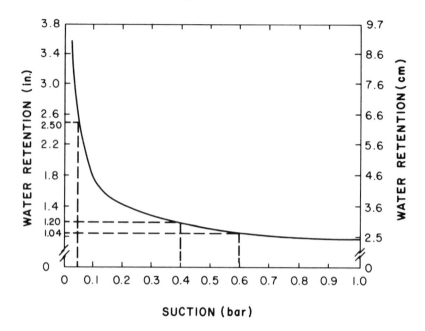

FIGURE 5.   Water retention in the zone of regulation (12 in. or 30 cm) of a Southern
Coastal Plain soil with 10 in. (25 cm) of loamy sand over sandy loam or fine sandy loam.

located along the radius indicated by the dash line. This will place five stations in soil type
NbB1 (55% of the area) and one station in type Nb (21% of the area). These locations could
be altered somewhat to allow for topographic variability. The tensiometers must be serviced
and read three times a week early in the season and more often as plant water demand
increases.

Begin irrigating the field when the main suction (average of all six tensiometer readings)
at the center of the water regulation zone — in this case, 6 in. (15 cm) — reaches 0.4 bar.
It is desired that no portion of the soil-water control zone should exceed a suction of 0.6
bar. How one operates the system to achieve this level of management will depend upon
the portion of the total field area that is irrigated in 1 day.

Consider the case when the full-circle travel time is set at 24 hr (0.3 in./day or 0.76 cm/
day net application). If, during peak evapotranspiration periods, irrigation is begun when
the mean suction reaches 0.4 bar, by the time the circle is completed (24 hr later), the
suction in the area irrigated last could be above 0.6 bar (Figure 5). To avoid this, irrigation
should begin at a mean suction value of about 0.20 bar. Consider the other extreme, when
the full-circle travel time is set at 96 hr (1.2 in. or 3.0 cm net application). Irrigation would
have to begin when the mean suction level in the first quadrant reached 0.08 bar if suctions
greater than 0.6 bar were to be avoided in the fourth quadrant. Applying 1.2 in. (3.0 cm)
of water to a soil with a suction of 0.08 bar would result in excessively deep percolation
below the soil-water control zone, and should be avoided. A system designed for a 4-day
full-circle travel time characteristically will either apply water prematurely at the beginning
of the cycle or too late at the end. As full-circle travel time is decreased, this problem is
diminished. A solution to the problem would be to change the depth of application in each
quadrant to that necessary to bring the suction back to 0.05. For example, the first quadrant
might receive 0.3 in. (0.76 cm), the second 0.6 in. (1.52 cm), the third 0.9 in. (2.28 cm),
and the fourth 1.2 in. (3.0 cm) by adjusting the system full-circle travel times to 6, 12, 18,
and 24 hr, respectively. If this is not practical, an alternate approach would be to reduce
the full-circle travel time to 2 days (0.6 in. or 1.52 cm, net application) and begin the initial

Table 9
NITROGEN (N) AND POTASSIUM (K)
REQUIREMENTS FOR CORN AT VARIOUS
GROWTH STAGES

	Nutrient requirements (% of total)	
Growth stage	**N**	**K**
Emergence to 4 fully emerged leaves	2	2
4 to 8 fully emerged leaves	5	5
8 to 12 fully emerged leaves	19	29
12 to 16 fully emerged leaves	26	33
16 fully emerged leaves to silking and pollen shed	12	21
Silking to 12 days after silking (blister stage)	12	10
12 to 24 days after silking (dough stage)	10	0
24 to 36 days after silking (early dent)	7	0
36 to 48 days after silking (full dent)	2	0
48 to 60 days after silking (maturity)	5	0

From Hanway, J. J., How A Corn Plant Develops, Iowa State Univ. Coop. Ext. Serv. Spec. Rep. 48, Iowa State University, Ames, 1971. With permission.

cycle when the suction in the first two quadrants (one half of the field) reached approximately 0.15 bar. This would result in an acceptable amount of overwatering at the beginning of the cycle and suctions not excessively above 0.6 bar before irrigation near the end of the cycle (end of 2nd day). After the initial cycle had been completed, and if no rainfall occurred, the next cycle would begin when the mean suction value along the nearest radius (in the direction of travel) reached approximately 0.40 bar. During this subsequent "non-start-up cycle," no segment of the circle should exceed 0.4-bar suction before it is irrigated. When significant (more than 0.5 in. or 1.25 cm) rainfall interrupted the cycle, it would then be necessary to return to the "start-up cycle" procedure or a modification thereof.

This "maximum demand" irrigation schedule will only apply during extended rain-free periods that occur at the time of peak evapotranspiration demand. Most of the time during the growing season, the daily water needs of the crop will be less. Consequently, it is extremely important to locate a second tensiometer at the bottom of the soil-water control zone to determine if the proper amount of water is being applied at each irrigation. About 6 to 12 hr after an irrigation, the bottom of the control zone should have water-suction values in the range of 0.08 to 0.15 bar. If the suction is less than 0.08 bar, less water should be applied during the next irrigation by reducing the full-circle rotation time (increasing travel speed). If the value is greater than 0.15 bar, more water should be applied by increasing the full-circle rotation time (reducing travel speed). This approach can be used for any full- or part-circle rotation time — 12 hr, 1 day, 2 days, and so forth.

Fertilizer applications through the irrigation system can boost yields and lower production costs. This is especially true for those nutrients that are highly leachable, such as nitrogen, and somewhat leachable, such as potassium. Frequent applications quantitatively adjust to the uptake rate of the corn are very desirable. Table 9 shows the seasonal uptake rate of nitrogen and potassium for corn in relation to growth stage. Assuming that 300 lb of nitrogen per acre (335 kg/ha) and 200 lb of potassium per acre (224 kg/ha) are required for a yield of 200 bushels/acre (12.54 metric ton/ha) the nitrogen and potassium requirement for each stage can be calculated. Although the time interval between applications will be partially dependent on weather conditions, a 7- to 14-day interval during rapid uptake periods may be necessary to insure the plant needs are met and serious leaching losses are avoided.

Alternative System

A cable-tow volume-gun system could be utilized as an alternate to the center-pivot sprinkler system. The selection would be predicated upon field shape, initial investment costs, and availability of labor. A typical cable-tow volume gun with a 1.75-in.-diameter (4.4-cm) nozzle at 85 lb/in.^{2} (586 kPa) will discharge 600 gal/min (37.8 ℓ/sec) over a wetted diameter of 460 ft (140 m). The application rate, AR, is calculated as shown below. It is a function of sprinkler discharge, Q_s, and area of wetted circle, A_w. The K-value is 96 for unbracketed units and 360 for bracketed.

$$AR = (K \times Q_s)/A_w$$

$$= (96 \times 600)/(230^2 \times 3.14)$$

$$= 0.35 \text{ in./hr } (0.9 \text{ cm/hr})$$

which is well below the maximum allowable rate of 0.7 in./hr (1.8 cm/hr) given in Table 2.

Assuming an application efficiency, E_a, of 75% and a lane spacing, S_L, of 290 ft (88 m); 63% of the wetted diameter for 10-mi./hr (4.5-m/sec) winds and a desired 1.2 in. (3.0 cm) of water applied; the travel speed required is calculated as follows:

$$V_t = (K \times Q_s \times E_a)/(S_L \times D)$$

$$= (1.605 \times 600 \times 0.75)/(290 \times 1.2)$$

$$= 2.08 \text{ ft/min } (0.63 \text{ m/min})$$

note that K = 1.605 for unbracketed units and 6.0 for bracketed.

Utilizing a 660-ft (201-m) flexible hose and irrigating beyond the end of the hose a distance equal to approximately one third the lane spacing, it is possible to irrigate a strip 1513 ft (461 m) long (2 × 600 + 0.67 × 290). Traveling at 2.08 ft/min (0.63 m/min), it will require 10.6 hr (2 × 660/2.08 × 60) to irrigate one complete strip. To irrigate 80 acres (32.5 ha) with strips 290 ft wide by 1513 ft long would require eight lanes. Allowing 1.4 hr of moving time between lanes, it would take 4 days (8 × 12/24) to irrigate the full 80 acres. Consequently, to irrigate 160 acres (64.8 ha) and stay within our maximum irrigation cycle period of 4.3 days would require a two-gun system.

Determining when to irrigate and if the correct amount of water is being applied is carried out in the same manner as for the center-pivot system. The only difference is in selecting the locations for the tensiometer stations. Use a minimum of four tensiometer stations per 80-acre (32.4-ha) block. If the stations are in alternate lanes, the volume gun would always be within 1 day of travel time of a station. Two tensiometers should be installed at each station; one at the midpoint of the soil-water control zone (6 in. or 15 cm) and a second at the bottom of the zone (12 in. or 30 cm). The location of the station within each lane should be based on soil type, topography, and accessibility. One possible field layout for the system including tensiometer stations is shown in Figure 6.

The ''start-up'' irrigation cycle should begin when the mean value of the four tensiometers (at the 6-in. depth) in each 80-acre area is approximately 0.15 bar. This will result in over-

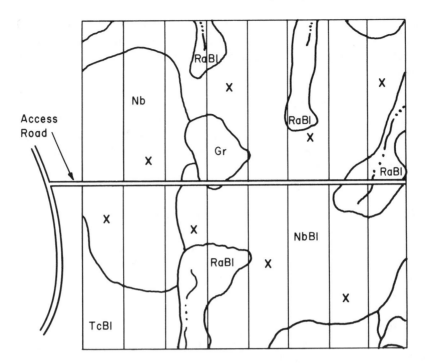

FIGURE 6. Proposed field layout for a 160-acre (65-ha) cable-tow volume gun (two units) irrigation system. X, tensiometer station; Gr, Grady sandy loam; Nb, Norfolk loamy sand, level thick surface phase; NbBl, Norfolk loamy sand, very gently sloping thick surface phase; TcBl, Tifton sandy loam, very gently sloping phase; RaBl, Rains loamy sand, very gently sloping thick surface phase.

irrigation of the first three or four strips and suctions greater than 0.6 bar in the final strip prior to irrigation. The other alternative is to reduce the depth of application (by increasing travel speed) to 0.3 in. (0.76 cm) for lanes 1 and 2, 0.6 in. (1.52 cm) for lanes 3 and 4, 0.9 in. (2.29 cm) for 5 and 6, and 1.2 in. (3.0 cm) for lanes 7 and 8. If no significant rainfall occurs, subsequent cycles should begin when the tensiometers in lane numer 1 (or 2) approach a mean value of 0.3-bar suction. If a cycle is interrupted by a significant amount of rainfall, the start-up procedure can be used when irrigation is called for. In this situation, lane number 1 becomes the next lane to be irrigated.

The amount of water applied should be such that the suction at the bottom of the soil-water control zone is in the 0.08- to 0.15-bar range 6 to 12 hr after an irrigation. If the mean suction consistently falls outside this range, the travel speed of the volume gun should be adjusted accordingly.

Tensiometers should be read and serviced a minimum of three times weekly during the early growing season and mild weather, and more frequently during the mid and late growing season and hot weather.

Peaches in Oconee County, Ga.

In Oconee County, a grower wishes to set out 24 acres (9.7 ha) of peaches on a Cecil sandy loam (Typic Hapludult) with 2 to 4% slope.[7] The sandy loam surface soil is about 7 in. (18 cm) deep over about 6 in. (15 cm) of sandy clay loam, which classifies it as soil-texture grouping B in Table 1. There is a moderate erosion hazard, and control measures must be imposed. Long-term weather records indicate about a 35% chance of receiving 1 in. (2.5 cm) of rainfall per week from early April through July, except from May 15 to June

15, when chances decrease to about 30%. From April through October there is more than a 75% chance that monthly evaporation will exceed rainfall. An adequate supply of water is available in a nearby lake — 15 acre-feet (18,505 m³) — or a well with a 75-gal/min (4.73 ℓ/sec) capacity.

Crop Culture and Water Regulation

Sodded middles with vegetation-free strips 5 to 6 ft (1.5 to 1.8 m) wide beneath the trees are recommended for peach culture. The strips should be kept clean with an effective herbicide. The middles should be laid out across the prevailing slope to reduce erosion and vehicular energy requirements. Trees will be planted on a 20 × 20 ft (6.1- by 6.1-m) spacing pattern to allow room for vehicles.

Drip irrigation is well suited to peach trees. Water can be applied daily at low pressures to a small area beneath each tree. This minimizes water and energy requirements while meeting the almost continuous water needs of the tree. A single lateral line will be laid along the clean strip at the base of the trees. One to six emitters, depending on tree size, attached to the lateral line will supply each tree's water requirements. Where surface damage by traffic is a problem, the lateral lines can be buried and a "riser outlet" provided for the emitters at each tree.

Sizing the System

Emitter capacity should be 1.0 gal/hr (3.8 ℓ/hr). For a medium-textured soil, this flow rate should prevent surface ponding and runoff from the wetted area beneath the emitter.

Under high-level irrigated management, peach trees can be expected to reach full canopy size of 15 to 18 ft (4.6 to 5.5 m) in diameter in 4 years after field transplanting. The peak evapotranspiration demand for a two thirds sod cover (Table 6) should be approximately 0.23 in./day (0.58 cm/day). The peak transpiration rate, T (inches or centimeters per day), for a mature tree with a canopy diameter of 16 ft (4.9 m) is calculated as follows:

$$T = (ET)(P_s/85)$$

$$= 0.23 \, (50/85)$$

$$= 0.13 \text{ in./day } (0.34 \text{ cm/day})$$

where P_s is the percentage of total area shaded by crop canopy and $P_s/85$ has a maximum value of one.

Assuming a transpiration ratio, TR, of 0.95 and an emission uniformity, EU, of 0.92, the peak water requirements per tree, Q_p, are calculated:

$$Q_p = (K \times T \times A)/(TR \times EU)$$

$$= (0.623 \times 0.13 \times 400)/(0.95 \times 0.92),$$

$$= 38 \text{ gal/day } (145 \text{ 1/day}).$$

where T equals transpiration rate; A equals gross area; K equals 0.623 for U.S. customary units and 10 for metric units.

Selecting the total area wetted by emitters to be 17% of the gross area, A_w, the number of emitters per tree, N_e, is calculated as follows, where the wetted area per emitter is obtained from Table 10.

$$N_e = A_w/A_e$$
$$= (0.17 \times 400)/13.2$$
$$= 5$$

Spacing the emitters at 4-ft (1.2-m) intervals along the lateral line will allow the wetted areas to just meet (wetted diameter = 4.1 ft or 1.25 m from Table 11). Emitters should not be arranged so that one is right next to the tree trunk. This will reduce soil support for the trunk because of the wetting of the soil. It is preferable to place an emitter 2.0 ft (0.6 m) on either side of the trunk. Because the trees are smaller when planted, no more than two emitters are needed the first year.

The net length of time, I_t, for applying the peak daily water requirement, when Q_e equals discharge per emitter, is calculated as follows:

$$I_t = Q_p/(N_e \times Q_e)$$
$$= 38/(5 \times 1.0)$$
$$= 7.6 \text{ hr}$$

To minimize system flow requirements and pump size, the system should operate continuously during peak demand periods. To achieve this, the field should be divided into a number of subunits, N_s, for a 24-hour-a-day operation as follows:

$$N_s = 24/I_t$$
$$= 24/7.6$$
$$= 3$$

The field layout of the system will then include three independently controlled subunits of 8 acres (3.2 ha) each.

The water requirements, Q_d, to meet the peak daily demand of the 24-acre (9.7-ha) orchard will depend on subunit area, A_s, number of plants per unit area, N_p, the peak water requirement per plant, Q_p, and length of water application, I_t, as follows:

$$Q_d = (N_p \times Q_p \times A_s)/(K \times I_t)$$
$$= (109 \times 38 \times 8)/(60 \times 7.6)$$
$$= 73 \text{ gal/min (4.6 } \ell\text{/sec)}$$

where K = 60 for U.S. customary units and 3600 for metric units

Operating the System

Determining when and how much to irrigate is our next consideration. A minimum of two tensiometer stations should be installed in each subunit. These stations should be located at easily accessible sites that represent the predominant soil and topography over the subunit area. Two tensiometers should be utilized to determine when to irrigate the subunit. One tensiometer should be installed at a depth of 8 in. (20 cm) below the emitter, one half the depth of the soil-water control zone shown in Table 3, and 12 in. (30 cm) horizontally (one

Table 10
AREAS WETTED BY EMITTERS ON COARSE, MEDIUM, AND FINE SOILS

Emitter discharge rate		Soil texture[a]																
		Coarse						Medium						Fine				
		Diameter		Area				Diameter		Area				Diameter		Area		
Gallons per hour	Liters per hour	Feet	Meters	Square feet	Square meters			Feet	Meters	Square feet	Square meters			Feet	Meters	Square feet	Square meters	
0.5	1.9	1.2	.37	1.2	0.11			2.9	0.88	6.5	0.60			4.1	1.25	13.2	1.23	
1.0	3.8	2.5	.76	4.7	0.44			4.1	1.25	13.2	1.23			5.4	1.65	22.5	2.09	
2.0	7.6	4.1	1.25	13.2	1.23			5.4	1.65	22.5	2.09			7.0	2.13	38.3	3.56	

[a]　Coarse — sands, loamy sands. Medium — very fine sandy loam, loam, silt loam, silt. Fine — sandy clay, silty clay, clay.

From Keller, Jack and Karmeli, David, Trickle Irrigation Design, Rain Bird Sprinkler Manufacturing Corp., Glendora, Calif., 1975. With permission.

fourth the wetted diameter) from the emitter. It is desirable to fix (by pinning) the emitter and lateral line to the soil surface to prevent any accidental change in the configuration. When the mean suction value reaches 0.30 bar, the subunit should be irrigated. This should prevent the suction from exceeding 0.60 bar by the time irrigation is complete, which could be up to 22 hr if all three subunits call for water at the same time during a peak demand period.

The second tensiometer at each station provides a means of determining if too little or too much water is being applied at each irrigation. It should be installed near the bottom of the control zone at a depth of 15 to 16 in. (38 to 41 cm), and located 12 in. (30 cm) horizontally from the emitter (one fourth the wetted diameter). Twelve to 24 hr after an irrigation, these tensiometers should read between 0.08- and 0.15-bar suction. If the value is consistently less than 0.8 bar, reduce the irrigation time and apply less water. If the value exceeds 0.15-bar suction, the irrigation time should be increased (up to a maximum of 8 hr during peak evapotranspiration demand).

Switching tensiometers could be installed in each subunit to automatically turn the water on when the suction reaches a preset value. These tensiometers should be located in the same position (relative to emitter) as the shallow unit in the manually read tensiometer stations. The tensiometer switch should be set to turn the system on at approximately 0.30-bar suction. The switches will normally close after a 0.04- to 0.06-bar drop. If a switching tensiometer is used, we strongly recommend frequent reading of tensiometers as a further check on the performance of the system.

The importance of regular service and reading of the tensiometers cannot be overemphasized. Three times a week is suggested as a minimum. During high water-demand periods, it may be necessary to read the tensiometers daily. Plotting these data will enable the irrigator to trace the history of the suction levels in the soil-water control zone and anticipate when and how much to irrigate. If switching tensiometers are used, the manually read data will help to determine if the switching tensiometer-emitter configuration is adequate.

If possible, a well should be used as a source of water for the drip system. Surface water supplies such as ponds, lakes, and rivers require an intensive filtration system which may not entirely prevent emitter clogging. Filtration requirements for well water can generally be satisfied with a 100-mesh (or smaller) screen-type or equivalent cartridge-type or sand filter.

REFERENCES

1. **Hagan, Robert M., Haise, Howard, R., and Edminster, Talcott, W.,** *Irrigation of Agricultural Lands,* American Society of Agronomy, Madison, Wis., 1967.
2. **Hanway, J. J.,** How A Corn Plant Develops, Iowa State Univ. Coop. Ext. Serv. Spec. Rep. 48, Iowa State University, Ames, 1971.
3. **Jensen, E. R., Aull, L. E., Shepard, J. L., Thomas, C. B., Carter, R. L., Haygood, E. S., and Middleton, R. G.,** Soil Survey, Tift County, Georgia. Soil Conservation Service, U.S. Department of Agriculture, Washington, D.C., 1959.
4. **Keller, Jack and Karmeli, David,** Trickle Irrigation Design, Rain Bird Sprinkler Manufacturing Corp., Glendora, Calif., 1975.

5. **Pair, Claude H., Hinze, Walter W., Reid, Crawford, and Frost, Kenneth R.,** *Sprinkler Irrigation,* 3rd ed., Sprinkler Irrigation Association, Silver Springs, Md., 1975, 337.
6. **Rhoads, F. M., Mansell, R. S., and Hammond, L. C.,** Influence of water and fertilizer management on yield and water input efficiency of corn, *Agron. J.,* 70, 305, 1978.
7. **Robertson, Stanley, M.,** Soil Survey, Clark and Oconee Counties, Georgia, Soil Conservation Service, U.S. Department of Agriculture, Washington, D.C., 1968.
8. **Stansell, J. R., Shepherd, J. L., Pallas, J. E., Bruce, R. R., Minton, R. A., Bell, D. K., and Morgan, L. W.,** Peanut responses to soil water variables in the Southeast, *Peanut Sci.,* 3, 44, 1976.
9. **Taylor, Sterling A. and Ashcroft, Gaylen L.,** *Physical Edaphology,* W. H. Freeman & Co., San Francisco, 1972.
10. Soil Improvement Committee, Western Fertilizer Handbook, California Fertilizer Association, Sacramento, 1961.

INFORMATION TO CONSIDER IN DESIGNING AND SELECTING IRRIGATION SYSTEMS FOR AGRICULTURE AND A DESCRIPTION OF DIFFERENT TYPES OF SYSTEMS

Ronald E. Sneed

INTRODUCTION

The history of recorded man's existence on this planet is filled with accounts of his involvement with irrigation. For many centuries, all irrigation systems were gravity systems where water was discharged from a ditch or pipe to the head of a field and water flowed by gravity down the length of the field. Factors such as water supply, soil type, field slope, crop, etc. were considered in designing the system. In recent decades, sprinkler and drip systems have been introduced, and design has become somewhat more complicated, but many of the same factors that were considered many centuries ago must still be considered in design and selection of irrigation systems today.

The factors that should be considered are

1. Type of irrigation system
2. Water supply
3. Soil type(s)
4. Crop(s)
5. Topography
6. Field(s) size and shape
7. Labor supply
8. Fuel source
9. Need for environmental modification
10. Chemigation
11. Rainfall, temperature, and wind speed
12. Number of hours of operation per day
13. Financial resources
14. Lease vs. purchase
15. Dealer availability

The order in which these factors are considered is not necessarily in the order given, but each should be considered.

The most important factor is the water supply. The system is of no value unless there is water available. For a reasonably efficient irrigation system, to apply 1 in. of effective irrigation to one acre of land requires between 33,000 and 37,000 gal of water. Water sources include lakes, impounded ponds, dug pits, canals, streams, and wells. The source of water is not too important, provided sufficient water is available to meet the needs of the crop. Many ponds and irrigation pits are supplied by surface runoff and have little or no ground-water recharge. These must store enough water to meet crop needs during the irrigation season. Stream flow will be less during drought periods. Wells will provide a relatively constant supply throughout the year. Ground-water resources vary greatly across the U.S., and in many areas, satisfactory irrigation wells cannot be developed. Well drillers and water resource officials in each state can provide data on ground water availability. Some growers are now going to combination water sources, such as a pond that is partially recharged by a well. Where the irrigation system is not operated continuously, the well capacity can be reduced, resulting in a lower well cost. The cost of a combination water supply system requiring two pumps and power units must be compared to the cost of a single water source and one pump and power unit. In many areas, adequate ground water may not be available to operate an irrigation system, but by combining surface storage and ground water, the water supply will be adequate.

Water supplies need to be developed in the off-irrigation season. Contractors may not be as busy during this time and can better schedule their work and can possibly reduce their fees. Normally, well construction costs will be lower if the contractor is not required to guarantee a certain pumping rate. Test wells can provide good data on potential yield from wells and are a good investment.

Many states have laws relating to development of water supplies for irrigation. For example, North Carolina requires a permit to drill an irrigation well and to construct a pond that impounds more than 10 acre-ft of water and where the dam is more than 15 ft high. Use of stream water in the eastern U.S., with a few exceptions, is governed by the doctrine of Riparian Rights. Owners of land adjacent to a stream can make reasonable use of the water for beneficial uses, but are expected to allow some downstream flow. In the western U.S., prior appropriation controls the use of stream water. The right to use water is appropriated to users, and water rights belong to the land.

Soil type is important for several reasons. Water application rates, total amount of water to be applied at one irrigation, and frequency of irrigation are affected by soil type. In fields that have more than one soil type, the system must be designed for the major soil type. Normally, sandy soils will have a higher intake rate, a lower water-holding capacity, and will require more frequent irrigation than will clay soils.

The amount of organic matter in the soil will affect the water intake rate and the water-holding capacity. Organic matter is very light, and even if it is saturated, it does not hold a great amount of water. However, rates up to 6 to 10% are beneficial in increasing water-holding capacity. Organic matter also improves water intake. Soil salinity reduces the availability of water to the plant. Soil structure can affect crop root development. Granular and crumb structures are the most desirable for plant growth. Good soil structure makes soil more permeable, allows excess water to drain more readily, and allows deeper root development, therefore increasing the effective water-holding capacity.

Chemical and physical barriers in soil can affect water-holding capacity. Low pH, high aluminum, and high sodium can affect availability of water and nutrients and restrict root development. Hardpans and high water tables can affect root development and restrict soil moisture availability.

Loam, silt loam, and silty clay loam soils have the greatest available water-holding capacities. Addition of organic matter can improve water-holding capacity. Chemical and physical barriers can reduce available soil moisture. The soil type on which crops are grown can affect the profitability of irrigation.

The crop to be irrigated will play a major role in the profitability of irrigation. Some crops such as tobacco, soybeans, peanuts, and hay crops have the potential to recover from short drought periods with minimal reduction in yield or quality. Other crops such as corn and most of the short season horticultural crops are drastically affected by droughts at the pollination and fruit development stage of growth. The yield response of some crops to irrigation may not be sufficient to return a profit after subtracting the cost of owning and operating an irrigation system. Growers should study all available data on crop response to irrigation. They should consult with neighbors that may be irrigating. Once a decision has been made to irrigate a particular crop, select varieties that respond best to irrigation. Do not neglect other cultural practices that could negate the effects of irrigation.

Rainfall distribution can significantly affect the need for irrigation. Most states have available information on historical rainfall distribution. Some states have published information on probability of drought days and average rainfall amounts. A potential irrigator should study this data to determine the extent of drought in his area. There is considerable variation in rainfall. Localized rainfall distribution may vary considerably from published data. Temperature and wind speed can affect the amount of irrigation required, the uniformity of application and the type of irrigation system used.

Once it has been determined that an adequate water supply is available or can be obtained and that it may be profitable to irrigate, one can then begin to decide on what type of irrigation system will best fit a particular situation. There are a number of people who can provide assistance. The Agricultural Extension Service and the Soil Conservation Service have personnel who specialize in irrigation. Consultants may be employed to provide design assistance. Irrigation dealers provide assistance on system selection, design, and operation. It is recommended that several irrigation dealers be contacted. Neighbors that are irrigating can be of assistance in equipment selection.

There is a wide choice of irrigation systems from which to select. They are not all equally adaptable to a particular situation. Each of the several systems is briefly described later in this chapter. Even though there are a number of types of irrigation systems, for most growers the choice is limited either because of initial cost per acre, topography, field size and shape, crop, soil type, acres to be irrigated, labor available, or whether the system will be used for purposes other than supplying soil moisture.

Purchasing an irrigation system means the purchase of a variety of components to complete a system. These components are provided by several manufacturers, and it is up to the dealer or a consultant to insure that all the components fit together to provide a system that will meet the desired needs, namely, to meet the peak moisture demands of the crop to be irrigated. This is an area in which many systems fail. Many systems are sold on the basis of supplying 1 in. of irrigation per week. This is not adequate for many crops. For example, corn during the period from pollination to dough stage of growth will use as much as 0.3 in. of water per day or more than 2 in./week. Other crops have similar moisture demands, especially during extended dry periods. Most irrigation dealers do a reasonably good job of designing to meet these peak demands; however, there are systems that are inadequate. This often occurs where one pump is being used to supply two center pivots, or a system is designed to irrigate two crops with different critical moisture periods under the pivot, and the grower plants the whole circle in one crop. Occasionally, there is no allowance made for evaporation and transmission losses.

With the cable-tow and hose drag travelers, many growers try to irrigate too many acres and do a poor job of meeting the moisture demands. Occasionally, dealers oversell the system. They may indicate the system can irrigate a certain number of acres per week and do not always state that it is based on applying 1 in./week. Often the system is operated fewer hours per day than the system was designed to operate.

In some cases, pipe size and/or pump size and power unit are inadequate. In sizing PVC plastic underground pipe, try to hold the velocity to 5 ft/sec or lower to reduce friction loss. High flow velocities can increase pumping costs and also the potential for pipe damage due to water hammer. If there is the possibility of expanding a system or adding another unit, this needs to be considered in the early stages of planning. Sizing pumps is important, but equally important is pump efficiency. Some pumps are more efficient than others. Normally their initial cost is higher, but savings in operating cost will more than compensate for the higher initial cost.

Selection of power units should receive careful consideration. One needs to consider initial cost, maintenance requirements, operating costs, and life of the unit, auxiliary costs such as demand charges and possibility of off-peak operation for electric power and fuel tanks for internal combustion engines, and availability of various fuels.

One also needs to consider other uses of the irrigation system besides supplying soil moisture. Two areas that are often considered are environmental modification[20] and chemigation. Certain types of sprinkler irrigation systems such as the solid-set and permanent systems can be used for frost/freeze protection during the spring and fall and evaporative cooling to reduce heat stress during the summer. Frost/freeze protection using sprinkler irrigation works on the latent heat of fusion principle that water undergoing a phase change from liquid to solid liberates heat, 80 cal/g. If sufficient water is being applied under a no wind or low wind condition, the temperature of the plant and plant parts can be maintained near 32°F (0°C) even though the air temperature outside the irrigated area may drop to 20 to 22°F (-5.6 to 6.7°C). Rates of water application will vary from 0.10 in./hr for low growing crops such as strawberries to 0.18 in./hr for tall crops such as apple trees.

Evaporative cooling using sprinkler irrigation is a method of reducing heat stress on plants by lowering temperatures, raising relative humidity, and reducing transpirational water losses. Low volumes of water are required. A continuous application of 0.06 in./hr or a higher rate applied in a sequencing on-off mode should be satisfactory.

Chemigation is a term that has been coined to describe the application of fertilizer and pesticides through irrigation systems. Threadgill[21] estimated that in 1983, there were approximately 4.3 million ha (10.6 million acres) of land chemigated. On this total, approximately 79% was fertigation, 14% was herbigation, 4% was insectigation, with the remaining being fungigation and nemigation. Of the total area chemigated, 84% was with sprinkler irrigation systems, 13% with surface irrigation systems, and 3% with trickle irrigation systems. Indications are good that the acreage being chemigated will continue to increase. The largest acreage will probably continue to be fertigation, but the acreage to which pesticides will be applied through the irrigation system will continue to grow.

Fertilizer and pesticides can be successfully applied through certain types of irrigation systems if the system is properly designed. Depending upon the material to be applied, the irrigation system must have the capability to apply varying amounts of water. Most insecticides and fungicides will need to be applied with 0.1 in. of water or less and the material needs to be applied uniformly to the foliage of the plants. Herbicides and nematicides should be applied uniformly to the soil surface. It is possible to apply fertilizer and selected pesticides with trickle and surface irrigation systems. The full range of pesticides and selected fertilizers can be applied with solid-set, permanent, center pivot, and linear-move sprinkler irrigation systems. To apply insecticides and fungicides, the center-pivot and linear-move systems need to be equipped with high speed gear boxes to facilitate rapid movement across the field and application of small amounts of water.

It may also be possible to apply fertilizer through cable-tow and hose drag self-propelled traveling gun sprinklers. Wright[22] has conducted limited studies on application of nitrogen fertilizer to grain corn using a hose drag traveler. This method of application appears to be successful.

Threadgill[21] reports that chemigation is less expensive than conventional ground or aerial application if more than one application of material is made to a field in 1 year. Threadgill references a number of researchers that have conducted work on chemigation. In general, the results of these studies indicate that chemigation is feasible and can be effective when the irrigation system and chemical injection system are properly installed and operated and when the proper chemical formulation and water application rates are used. Where foliar retention of the chemical is required for efficacy, chemical formulation appears to be very important. For foliar applications, the most consistent positive results have been obtained when the technical chemical has been formulated in an oil without the addition of any emulsifier.

In addition to the importance of uniform application and positive results, it is necessary to insure that the chemicals not contaminate the water supply. The American Society of Agricultural Engineers[10] has developed an engineering practice (ASAE EP 409), Safety Devices for Applying Liquid Chemicals Through Irrigation Systems. A number of states have regulations or proposed regulations relating to proper safety controls to use on an irrigation system that is used to apply chemicals. Safety controls include check valves, antisiphon valves, low pressure drain valves, and interlocked chemical feed pumps and irrigation pumps.

The last items that should be considered are selection of a dealer, making sure that there is adequate time to design and install the system before it is needed, and whether to lease or purchase the equipment. Of the sprinkler systems, the center pivot and the lateral move require the longer time between the decision to purchase and placing the system in operation.

There is a variety of equipment on the market and a number of people selling equipment. Select a dealer who has a reputation for good design and service. All of this equipment will require maintenance, and if it fails during a drought period, service is needed within a few hours. Attend irrigation meetings, talk to other farmers, visit systems in operation. Find out as much about the equipment as possible before a decision is made to purchase. Do not purchase based solely on initial price. Examine the cost of operating the equipment. For large mechanical move systems, such as the center-pivot and linear move, try to purchase from a local reputable dealer. Most of these machines are too complex to be repaired by the owner.

More farmers are leasing irrigation equipment. Only the purchaser and the accountant can decide if it is the correct decision for your operation. Look at both purchase and lease and see which best fits your financial situation. Some manufacturers offer leasing plans that normally have a slightly lower annual interest rate than commercial leasing companies. Normally, the total cost will be greater for lease than for purchase.

Purchasing an irrigation system is a major decision. With proper planning, it can be a wise decision for many growers.

DESCRIPTION OF DIFFERENT TYPES OF IRRIGATION SYSTEM

Portable Aluminum Pipe Sprinkler System

This is an aluminum pipe system designed to cover at one time only a portion of the acreage to be irrigated. The system is moved one or more times per day. It may use a variety of sprinkler sizes from small sprinklers to gun sprinklers. The system is adaptable to most soil types, topography, field sizes and shapes. It has the highest labor requirement of any sprinkler system. Initial cost per acre is medium with medium-low to medium-high operating pressure, medium to medium-high operating cost, low to medium-high water application rate, and excellent to fair uniformity of application. As the sprinkler size increases, the labor requirement is reduced, but normally because of labor requirements the system is used on small acreage. The initial cost of this system will vary from 400 to $800 per acre based on 1986 prices. Initial cost per acre for all the systems listed are based on 1986 prices.

Solid-Set Irrigation System

This is an aluminum or polyvinyl chloride (PVC) plastic pipe system that is placed in the field(s) at the start of the irrigation system and left in place throughout the season. It may use a variety of sprinkler sizes from small sprinklers to gun sprinklers, but the system is normally used with small sprinklers to provide a low application rate. It is adapted to almost all soil types, topography, field sizes and shapes. The initial cost per acre is very high with low labor requirement and medium-low to medium-high operating cost. Uniformity of water application will vary from excellent to fair depending on sprinkler spacing and size. With small sprinklers it is excellent for environmental modification[20] and chemigation. The system can be automated. Normally, it is used for crops that have a high cash value. The initial cost will vary from 1500 to $3000 per acre.

Permanent Sprinkler System

This is an underground pipe system with only a portion of the risers and the sprinklers above ground. This system has essentially the same characteristics as the solid-set system, except the labor requirements will be lower and the system cannot easily be moved to another location. The initial cost per acre will be in the same price range as the solid-set system. In some parts of the U.S., the solid-set and permanent system is referred to by the same name.

Side-Roll Wheel Move System

This is a wheel move aluminum lateral line system where the lateral line is the axle for the wheels. It is designed to be used on rectangular or square fields up to 2000 ft in length and on crops 4 ft or less in height. The aluminum lateral line is attached to a portable aluminum or permanent supply line by a flexible pipe. In operation, the lateral line is stationary. Once a section has been irrigated, a small gasoline engine attached to the aluminum lateral line through drive wheels is started and the unit is moved to the next irrigation set, normally a distance of 60 ft. There are two systems available. One unit is automated, in that five moves can be made without disconnecting the lateral from the main line. In the automatic system, the irrigation pump is shut down, the lateral line is drained, the gasoline engine is started to move the unit to the next irrigation set, and the irrigation pump is started to place the system back in an irrigation mode. At the end of the fifth move, the flexible supply line is repositioned to another hydrant. This system is available from at least one manufacturer. The other systems are manual in that the irrigation pump must be stopped, the flexible pipe disconnected from the main line, the lateral drained and moved, then connected back to the main line and the irrigation pump started again. Normally, the wheels and sprinklers are spaced 40 ft apart. The system is limited to reasonably flat terrain. Uniformity of water application will vary from excellent to good. Medium pressure sprinklers are used, the initial cost per acre is medium, the operating cost is medium, the application rate is medium, and the labor requirement is low. The system has limited application since it requires a flat rectangular or square field of 20 acres or larger that has rows 1000 ft or longer. To move the system from field to field is difficult.

There is a variation of this system that has trail tubes behind the lateral line. Up to three sprinklers can be placed on each trail tube. A wider strip can be irrigated at each set. On this system, the main aluminum lateral is mounted on "A" frames with two wheels at each "A" tower, and crops up to 9 ft in height can be irrigated. This system can be end towed from field to field by rotating the wheels 90°. Side-roll wheel move systems have been popular in areas that produce alfalfa, sugar beets, potatoes, and other selected vegetables. Initial cost per acre will vary from 400 to $800 per acre.

Cable-Tow Traveler

This is a self-propelled, continuous move sprinkler system that utilizes a single sprinkler mounted on a two-, three-, or four-wheel trailer with water being supplied through a flexible rubber hose. The unit follows a steel cable that has one end anchored at the end of the field and the other end attached to the machine. The cable winds onto a cable drum on the machine as the machine moves through the field. Power to move the machine is supplied by a water turbine, water piston, water propeller, or an auxiliary internal combustion engine that drives the trailer direct or powers a hydraulic pump that then drives the unit. There are a variety of sizes of cable-tow travelers ranging from machines with capacities of less than 100 gal/min to more than 600 gal/min. Hose size varies from 2.5 to 4.5 in. and lengths from 330 to 1320 ft. There is also a small turf traveler that uses a 1-in. or 1 1/4-in. hose with lengths up to 200 ft. The cable-tow traveler is best suited to square or rectangular fields where the field length is twice the hose length, straight rows, flat to slightly rolling topography, and medium-high to high infiltration rate soils. This unit has a medium per acre initial cost, a high operating pressure, a high operating cost, and a medium-high to high water application rate. Uniformity of water application will vary from excellent to fair. Labor requirement is low. It is easily transportable from field to field and farm to farm. The initial cost of this system will vary from 425 to $800 per acre.

Hose Pull or Drum Traveler

The unit consists of a drum mounted on a trailer or wagon, a polyethylene hose, and a sprinkler cart on which is mounted a single gun sprinkler. The hose is connected to the drum and to the sprinkler cart and is used to supply water to the sprinkler and also to convey the sprinkler cart to the drum. The drum is rotated by a water turbine, water bellows, water piston, or an auxiliary internal combustion engine. As the drum is rotated, the hose is wound onto the drum and the sprinkler cart is drawn toward the drum. Hose sizes vary from 1 5/8 to 5 in. and in lengths from approximately 400 to 1320 ft. The unit is best suited to square or rectangular fields where the length is the same or double the hose length, straight rows, flat to slightly rolling topography, and medium-high to high infiltration rate soils.

The sprinkler cart will follow a contour, and this makes the drum traveler more adapted to contour farming than is the cable-tow traveler. The drum traveler has a medium to medium-high per acre initial cost, a very high operating pressure, high operating cost, medium-high to high water application rate and low labor requirement. Uniformity of water application will range from excellent to fair. It is easily transportable. With both the cable-tow and hose pull travelers, speed compensation should be a standard feature to provide for uniform travel speed through the field and as uniform application of water as possible. One advantage of the drum traveler is that only the amount of hose that is needed is unwound from the drum, and odd shaped fields can be irrigated. The initial cost of the hose pull traveler is 450 to $900 per acre.

Center-Pivot System

This is an electrically or hydraulically powered self-propelled lateral line on which sprinklers or spray nozzles are mounted. The system rotates around a center pivot point. The lateral line is supported by towers that are spaced 110 to 180 ft apart. The spacing on the towers is influenced by the size of the lateral line pipe, the land slope, and the size of the area being irrigated. Center pivots are available as fixed pivot point machines and as towable machines that can be used on two or more pivot points. For the towable machines, it is necessary to be able to tow the machine in a straight line from one pivot point to another pivot point. The pattern irrigated is a circle; however, with an end gun, it is possible to irrigate corners and odd shapes of a field. In some instances, it may be desirable to purchase a machine with a corner attachment that allows larger corners or odd shapes of fields to be irrigated.

Center pivots systems are best suited to square fields that are 40 acres or larger in size, slopes of less than 10% and medium-high to high infiltration rate soils. With center pivots, there is economy of scale. As the size of the system increases, the cost per acre will normally decrease. Center pivots are available as medium to high pressure, low pressure, and very low pressure systems. Because of the potential savings in energy, the trend is toward very low pressure systems. The very low pressure systems are best suited to high infiltration rate soils and flat terrain. There are a number of types of water distribution devices. Conventional rotary impact sprinklers are used on the medium to high pressure systems, controlled droplet size rotary impact sprinklers or spray nozzles are used on the low pressure systems, and spray nozzles of different types are used on the very low pressure systems. To reduce application rates, some systems use spray booms that spread the water over a large area.

Center pivots are available that are constructed of galvanized steel pipe, painted steel pipe, and aluminum pipe. The electric drive systems use three phase 440 V electric motors at each tower. The per acre initial cost of center pivots is medium-high to medium, operating pressure is medium to low, operating cost is medium to low, and labor requirements are very low. Uniformity of water application will vary from excellent to good. Center pivots have to be carefully designed to fit the area on which they are used. They are shipped directly to the user from the manufacturer. In most instances, site preparation is required to fit the system to the field. The initial cost of a center pivot will vary from 400 to $900 per acre with the larger systems having a lower per acre initial cost.

Linear-Move System

This is a self-propelled, electric drive, continuous move lateral line on which low or very low pressure spray nozzles or rotary impact or gear drive sprinklers are mounted. The system moves in a straight line down or across a field. The galvanized steel or painted steel lateral line is supported by towers spaced 110 to 180 ft apart.

Water is fed to the machine from an open ditch, a buried pipe line with hydrants and a flexible hose, or from a buried pipe line with hydrants where the machine automatically couples and uncouples itself from the hydrants. The boss tower, which controls the speed of travel, is located adjacent to the canal or buried pipe line. It follows a buried guidance wire or an above ground guidance cable. The boss tower can be located at the end of the lateral line or at some point in the lateral line. The linear move system is designed to be used on flat fields with widths up to 4000 ft and lengths up to 9000 ft. A lateral length to field length ratio of 1:2 up to 1:4 is desirable to reduce the per acre initial cost. Because the machine uses spray nozzles or low pressure sprinklers, the application rate is high, and it should be used on high infiltration rate soils. The per acre initial cost is medium-high to medium, operating pressure is low, operating cost is low, and labor requirements are very low. Uniformity of water application will vary from excellent to good. As with the center pivot, the linear move has to be carefully designed to fit the field, and some site preparation may be needed to match the system to the field. Both the center-pivot and the linear-move machines lend themselves to large acreage irrigation. Both are well adapted to chemigation. However, if a grower plans to apply insecticides and fungicides, high speed gear boxes are needed to allow the machine to move across the field at a rapid rate to minimize the total application of water. Some manufacturers provide a machine that can function as both a center pivot and a linear move. It moves as a linear down a field, then can be rotated near the end of the field in a semicircle and then moves as a linear down the other side of the field. The initial cost of a linear move will vary from 500 to $1000 per acre.

Drip or Trickle Irrigation

This is a low pressure system where a lateral line is placed down every crop row adjacent to the plant. Some growers place two crop rows on one bed and provide water to both rows from one lateral. Water is discharged from this line through emitters or orifices in the pipe or through micro-sprinklers. The lateral line can be buried adjacent to the plant row or may be placed on the soil surface. Normally, the emitter system is used for tree, shrub, and vine crops, and the pipe-orifice system is used for row crops. Emitters are available as standard and as pressure compensating. Most of the standard emitters are designed to operate in the range of 15 lb/in.^{2} The pressure compensating emitters will operate over a wider pressure range and still discharge a relatively uniform rate. Most of the pipe-orifice systems are designed to operate in the 6 to 12 lb/in.^{2} pressure range. Application rates will vary from 5 to 100 gal/min/acre depending upon the crop and the type of drip system being used. Drip systems can be used on most soil types. They work best on flat terrain, but can be used on slopes where rows are on the contour, and pressure regulating valves are used in the supply line to provide fairly uniform pressure on each lateral. The heart of the drip system is the filter system, which removes impurities from the water that might clog the emitters or orifices. Normally, less total water will be required for a drip system than for a sprinkler system. A number of companies also offer a line of micro-sprinklers. These may be rotary or stationary heads. Radius of throw is 3 to 20 ft or more. Pressure requirements are 10 to 35 lb/in.^{2} There is some question as to whether this is a drip system. Orifice size is larger than with conventional drip systems, and filtration requirements are reduced. Most of these systems are being used on tree crops, but some are being used on row crops and seed beds. Drip systems have a medium per acre initial cost, low operating pressure, low operating cost, medium to low application rate, and low labor requirements. Uniformity of water application varies from excellent to fair. For row crops using the pipe-orifice system, the in-field laterals are normally used 1 year and replaced. However, there are some row crop growers who are experimenting with using in-field laterals for several years either by using buried laterals or surface laterals that are retrieved at the end of the growing season. The drip system will have an initial cost of 400 to $1000 per acre and where the laterals are replaced annually, a replacement cost of 150 to $200 per acre.

Subirrigation

This is the maintenance of a water table at some depth below the soil surface through the use of ditches, subsurface drain tubing or tile, or mole drains and water control surfaces in ditches. The system is limited to flat terrain, soils that have an impermeable layer or permanent water table at a shallow depth of 4 to 8 ft, a sand layer at a depth of 3 to 4 ft, and a fairly permeable soil profile above the sand layer. Evans and Skaggs[4] state that the principle of the subirrigation system is to raise the water table in the soil by pumping water through the subsurface drain tubing, mole drains or field ditches, and to maintain that water level at a proper depth below the soil surface to supply demands of the crops for soil moisture. The water table may be allowed to fluctuate or be maintained at a fairly constant level.[19] Additional research is required to determine what is most desirable from a crop and water conservation standpoint. The subirrigation system also functions as a drainage system. If subsurface drain tubing is used, spacing of drain lines will normally be in the range of 50 to 120 ft. However, before a system is installed, hydraulic conductivity measurements should be made in the field to determine the rate at which water will move through the soil. Where this is not possible, a spacing of 60 to 70% of the recommended spacing for drainage can be used as a guide. A water supply of approximately 5 gal/min/acre[5] is recommended to meet the water demands of crops. Provisions need to be made to rapidly lower the water table during periods of rainfall. A subirrigation system is not recommended if the field does not need drainage.[3] A subirrigation system has a medium to high per acre initial cost, very low operating pressure, very low operating cost, and very low labor requirements.[5] Uniformity of water application will vary from excellent to fair. The initial cost of the system will vary from 400 to $1200 per acre.

Graded Furrow Surface Irrigation

This is the application of water in furrows that have a continuous slope in the direction of water movement to irrigate row crops. Row slopes should be in the range of 0.1 to 0.5%. Row lengths can vary from a few hundred to 1500 ft. The system is limited to fields that have natural slopes of 2% or less, but if the natural slope is in the 0.1 to 0.5% range, extensive land forming will not be required. It is best suited to soils that have low to medium infiltration rates. Fields must be well graded to provide uniform slope and as uniform water distribution as possible. Facilities should be installed to catch water that runs out of the lower end of the field. A graded furrow system is not designed to apply small applications of water. Normally, 2 or more in. of water will be applied per application. Water is applied to the upper end of the furrows by gated pipe or from ditches. Labor requirements are fairly high, initial cost per acre is medium, but will be dependent upon the amount of land forming required, operating cost is low, and operating pressure is low. Uniformity of water application will vary from good to fair. The initial cost of a graded furrow surface irrigation system will depend upon the amount of land grading required to prepare the field. Other considerations are whether water will be distributed through ditches and siphon tubes or through gated pipe. Initial cost can vary from 300 to $700 per acre.

Several authors[8,12-15] describe a system to automate and improve water distribution for graded furrow systems. Gates and Clyma[7] describe a method for designing furrow irrigation systems to improve seasonal performance.

Contour Levee Irrigation

Water is applied to nearly level strips or areas of land between two levees or dikes at a rate that exceeds the intake rate of the soil so that the area is rapidly covered.[18] This method has been used for many years to flood rice fields. It has been adapted to the irrigation of pasture, hay crops, and some row crops such as cotton, corn, and soybeans that are not damaged by temporary flooding or flushing. It is adapted to medium to fine textured soils, with a water-holding capacity in the root zone of at least 2.5 in. Slopes should not exceed 0.5%. On row crops, the levees are normally temporary. Some land grading will be required on most fields, but not to the extent often required for graded furrow surface irrigation.

Border Irrigation

Water is applied between two parallel dikes or border ridges, and each strip of land is irrigated independently. This system is normally used to irrigate flat planted crops in the western U.S. The dikes are normally 30 to 50 ft apart. The strip width should accommodate the farm machinery used for planting, tillage, and harvesting. It is best suited to slopes between 0.1 and 0.5% and soils with a medium infiltration rate. Water is applied at the upper end of the strip and allowed to rapidly spread over the entire width of the strip. The initial cost of this system will depend on the amount of land forming required and will vary from 250 to $650 per acre.

Basin Irrigation

Water is applied to an area of land surrounded by dikes or ridges. Both graded and level basins can be used; however, with the development of laser grade control land grading equipment, level basins have become more feasible.[2] This system is normally used to irrigate hay crops, but can be used to irrigate other flat planted crops. Water is applied at one end of the field (upper end on a graded basin) at a rapid rate and quickly moved across the entire field.[1] This system works best for level basins. Medium textured soils are more adapted to this method of irrigation. The width of the basin will depend on the water supply[16] and machinery used for planting, tillage, and harvesting. The initial cost of the system will depend upon the amount of land leveling required and can vary from 250 to $700 per acre.

Wild Flooding

This method of irrigation has been used for many centuries and is still used in many parts of the world. Water is allowed to flow out of a ditch or gated pipe across sloping land that has not been precision land graded. It is best adapted to irrigation of pasture and hay crops. Normally it is not an efficient method of irrigation; however, land grading costs are minimized. This system can be installed at an initial cost as low as $50 per acre and could cost up to $200 or more per acre.

FUTURE OF IRRIGATION

From 1964 to 1984 the irrigated acreage in the U.S. increased from approximately 41.6

million acres[17] to 61.6 million acres.[9] In that same period, the sprinkler irrigated acreage increased from 5.5 million acres to 22.2 million acres. The irrigated acreage in the 17 western states increased from 37.4 million to 52.0 million acres. Two things are significant from these figures. The sprinkler irrigated acreage grew much more rapidly than the surface irrigated acreage, and the irrigated acreage in the eastern states has grown more rapidly than the irrigated acreage in the western states.

These trends are expected to continue. The increase in the sprinkler irrigated acreage has occurred because of the availability of the mechanical-move systems (center-pivot, linear-move, cable-tow and hose pull travelers, and the side-roll wheel move). These systems have made irrigation of large areas feasible from the standpoint of economics and labor. Sprinkler irrigation systems are normally more efficient from the standpoint of water application than surface systems, but do require more energy to operate. There is a trend toward lower pressure sprinkler systems, and this will continue.

Most of the irrigation in the eastern states is in the southeast. This area is characterized by warm weather, yearly short to medium length drought periods, shallow crop rooting depths, soils with limited water-holding capacity, limited to adequate water supplies and crops that respond to irrigation. The irrigated acreage in southeastern U.S. will continue to increase. The rate of increase will be affected by the price of farm commodities, interest rates for money, and frequency of droughts.

Trickle irrigation, which is still in its infancy, will increase in popularity. Most of the trickle irrigation is now used on fruit and vegetable crops. However, trickle irrigation has been tested on cotton, sugar cane, peanuts, corn, and other crops.[11]

The total irrigated acreage in the U.S. will continue to increase. The availability of water, the price of farm products, and the cost of money and fuel or electric power will determine the rate of increase of irrigation. Due to shortage of water, some areas will not increase irrigated acreage and may see the irrigated acreage decline. Rising fuel costs have slowed the growth or in some cases reduced the irrigated acreage. Low commodity prices and the general farm economy have definitely slowed the growth of agricultural irrigation since 1984.

REFERENCES

1. **Clemmens, A. J.,** Depths of flow in level basins, *Trans. ASAE,* 23(4), 910, 1980.
2. **Erie, L. J. and Dedrick, A. R.,** Level-Basin Irrigation: a Method for Conserving Water and Labor, Farmers Bull. 2261, U.S. Department of Agriculture, Washington, D.C., S.E.A., 1979.
3. **Evans, R. and Skaggs, W.,** Agricultural Water Management for Coastal Plain Soils, AG-355, North Carolina Agricultural Extension Service, 1985.
4. **Evans, R. and Skaggs, W.,** Operating Controlled Drainage and Sub-Irrigation Systems, AG-356, North Carolina Agricultural Extension Service, 1985.
5. **Evans, R. O., Skaggs, R. W., and Sneed, R. E.,** Economics Evaluation of Controlled Drainage and Sub-Irrigation Systems, North Carolina Agricultural Extension Service, in press, 1986.
6. **Evans, R. O., Sneed, R. E., and Skaggs, R. W.,** Water Supplies and Control Structures for Controlled Drainage and Sub-Irrigation Systems, North Carolina Agricultural Extension Service, in press, 1986.
7. **Gates, T. K. and Clyma, W.,** Designing furrow-irrigation systems for improved seasonal performance, *Trans. ASAE,* 26(6), 1817, 1984.
8. **Goel, M. C., Kemper, W. D., Worstell, R., and Bondurant, J.,** Cablegation. III. Field assessment of performance, *Trans. ASAE,* 25(5), 1304, 1982.
9. **Goldstein, A.,** 1984 Irrigation survey, *Irrig. J.,* 35(1), 25, 1985.
10. **Hahn, R. H., Purschwitz, M. A., and Rosentreter, E. E.,** Safety devices for applying liquid chemicals through irrigation systems, ASAE EP 409, *ASAE Standards 1985,* 32nd ed., American Society of Agricultural Engineers, St. Joseph, Mich., 1985, 520.
11. **Hall, B. J.,** History of drip/trickle irrigation. Drip/trickle in action, Proc. 3rd Int. Drip/Trickle Irrig. Congr., ASAE, 1, 1, 1985.

12. **Kemper, W. D., Heinemann, W. H., Kincaid, D. C., and Worstell, R. V.,** Cablegation. I. Cable controlled plugs in perforated supply pipe for automatic furrow irrigation, *Trans. ASAE,* 24(6), 1526, 1981.

13. **Kincaid, D. C. and Kemper, W. D.,** Cablegation. II. Simulation and design of the moving-plug gated pipe irrigation system, *Trans. ASAE,* 25(2), 388, 1982.

14. **Kincaid, D. C. and Kemper W. D.,** Cablegation. IV. The by-pass method and cutoff outlets to improve water distribution, *Trans. ASAE,* 27(3), 762, 1984.

15. **Kincaid, D. C.,** Cablegation. V. Dimensionless design relationships, *Trans. ASAE,* 27(3), 769, 1984.

16. **Kruse, E. G.,** Lateral sizes for efficient level-basin irrigation, *Trans. ASAE,* 24(4), 961, 1981.

17. **Peace, H. L.,** Irrigation survey, *Irrig. Eng. Maint.,* 15(5), 9, 1965.

18. **SCS-USDA,** Contour-Levee Irrigation, SCS National Engineering Handbook — Irrigation, Sec. 15, Soil Conservation Service, U.S. Department of Agriculture, Washington, D.C., 1969, chap. 6.

19. **Skaggs, R. W.,** Water movement factors important in the design and operation of sub-irrigation systems, *Trans. ASAE,* 24(6), 1553, 1981.

20. **Sneed, R. E. and Unrath, C. R.,** Irrigation for Apple Orchards, AG-306, North Carolina Agricultural Extension Service, 1984.

21. **Threadgill, E. D.,** Chemigation via sprinkler irrigation: current status and future development, *Appl. Eng. Agric.,* 1(1), 16, 1985.

22. **Wright, F. S.,** ARS-USDA, Tidewater Research Center, Suffolk, Virginia. Personal correspondence, 1986.

Appendix

PREFERRED UNITS FOR EXPRESSING PHYSICAL QUANTITIES (AND THE CONVERSION FACTORS)

ASAE Engineering Practice EP285.6

USE OF SI (METRIC) UNITS*

SECTION 1 — PURPOSE AND SCOPE

1.1 This Engineering Practice is intended as a guide for uniformity incorporating the international System of Units (SI). It is intended for use in implementing ASAE policy, "Use of SI Units in ASAE Standards, Engineering Practices, and Data." This Engineering Practice includes a list of preferred units and conversion factors.

SECTION 2 — SI UNITS OF MEASURE

2.1 SI consists of seven base units, two supplementary units, a series of derived units consistent with the base and supplementary units. There is also a series of approved prefixes for the formation of multiples and submultiples of the various units. A number of derived units are listed in paragraph 2.1.3 including those with special names. Additional derived units without special names are formed as needed from base units or other derived units, or both.

2.1.1 Base and supplementary units. For definitions refer to International Organization for Standardization ISO 1000, SI Units and Recommendations for the Use of Their Multiples and of Certain Other Units.

Base Units
    Meter (m) — unit of length
    Second (s) — unit of time
    Kilogram (kg) — unit of mass
    Kelvin (K) — unit of thermodynamic temperature
    Ampere (A) — unit of electric current
    Candela (cd) — luminous intensity
    Mole (mol) — the amount of a substance
Supplementary units
    Radian (rad) — plane angle
    Steradian (sr) — solid angle

2.1.2 SI unit prefixes

Multiples and submultiples	Prefix	SI Symbol
10^{18}	exa	E
10^{15}	peta	P

---

* Prepared under the general direction of ASAE Committee on Standards (T-1); reviewed and approved by ASAE division standardizing committees: Power and Machinery Division Technical Committee (PM-03), Structures and Environment Division Technical Committee (SE-03), Electric Power and Processing Division Technical Committee (EPP-03), Soil and Water Division Standards Committee (SW-03), and Education and Research Division Steering Committee; adopted by ASAE December 1964; reconfirmed for one year, December 1969, December 1970, December 1971, December 1972; revised by the Metric Policy Subcommittee December 1973; revised March 1976; revised and reclassified as an Engineering Practice, April 1977; revised April 1979, revised editorially December 1979; revised September 1980, February 1982; revised editorially January 1985.

10^{12}	tera	T
10^{9}	giga	G
10^{6}	mega	M
10^{3}	kilo	k
10^{2}	hecto	h
10^{1}	deka	da
10^{-1}	deci	d
10^{-2}	centi	c
10^{-3}	milli	m
10^{-6}	micro	μ
10^{-9}	nano	n
10^{-12}	pico	p
10^{-15}	femto	f
10^{-18}	atto	a

2.1.3   Derived units are combinations of based units or other derived units as needed to describe physical properties, for example, acceleration. Some derived units are given special names; others are expressed in the appropriate combination of SI units. Some currently defined derived units are tabulated in Table 1.

SECTION 3 — RULES FOR SI USAGE

3.1   General. The established SI units (base, supplementary, derived, and combinations thereof with appropriate multiple or submultiple prefixes) should be used as indicated in this section.

3.2   Application of prefixes. The prefixes given in paragraph 2.1.2 should be used to indicate orders of magnitude, thus eliminating insignificant digits and decimals, and providing a convenient substitute for writing powers of 10 as generally preferred in computation. For example:

12 300 m or 12.3×10^{3} m becomes 12.3 km, and

0.0123 mA or 12.3×10^{16} A becomes 12.3 μA

It is preferable to apply prefixes to the numerator of compound units, except when using kilogram (kg) in the denominator, since it is a base unit of SI and should be used in preference to the gram. For example:

Use 200 J/kg, not 2 dJ/g

With SI units higher order such as m² or m³, the prefix is also raised to the same order. For example:

mm² is $(10^{-3}$ m$)^{2}$ or 10^{-6} m²

3.3   Selection of prefix. When expressing a quantity by a numerical value and a unit, a prefix should be chosen so that the numerical value preferably lies between 0.1 and 1000, except where certain multiples and submultiples have been agreed for particular use. The same unit, multiple, or submultiple should be used in tables even though the series may exceed the preferred range of 0.1 to 1000. Double prefixes and hyphenated prefixes should not be used. For example:

use GW (gigawatt) and kMW

3.4   Capitalization. Symbols for SI units are only capitalized when the unit is derived from a proper name; for example, N for Isaac Newton (except liter, L). Unabbreviated units are not capitalized; for example kelvin and newton. Numerical prefixes given in paragraph 2.1.2 and their symbols are not capitalized; except for the symbols M (mega), G (giga), T (tera), P (peta), and E (exa).

3.5   Plurals. Unabbreviated SI units form their plurals in the usual manner. SI symbols are

Table 1
DERIVED UNITS

Quantity	Unit	SI Symbol	Formula
Acceleration	Meter per second squared	—	m/s^2
Activity (of a radioactive source)	Disintegration per second	—	(disintegration)/s
Angular acceleration	Radian per second squared	—	rad/s^2
Angular velocity	Radian per second	—	rad/s
Area	Square meter	—	m^2
Density	Kilogram per cubic meter	—	kg/m^3
Electrical capacitance	Farad	F	$A{\cdot}s/V$
Electrical conductance	Siemens	S	A/V
Electrical field strength	Volt per meter	—	V/m
Electrical inductance	Henry	H	$V{\cdot}s/A$
Electrical potential difference	Volt	V	W/A
Electrical resistance	Ohm	Ω	V/A
Electromotive force	Volt	V	W/A
Energy	Joule	J	$N{\cdot}m$
Entropy	Joule per kelvin	—	J/K
Force	Newton	N	$kg{\cdot}m/s^2$
Frequency	Hertz	Hz	(cycle)/s
Illuminance	Lux	lx	lm/m^2
Luminance	Candela per square meter	—	cd/m^2
Luminous flux	Lumen	lm	$cd{\cdot}sr$
Magnetic field strength	Ampere per meter	—	A/m
Magnetic flux	Weber	Wb	$V{\cdot}s$
Magnetic flux density	Tesla	T	Wb/m^2
Magnetomotive force	Ampere	A	—
Power	Watt	W	J/s
Pressure	Pascal	Pa	N/m^2
Quantity of electricity	Coulomb	C	$A{\cdot}s$
Quantity of heat	Joule	J	$N{\cdot}m$
Radiant intensity	Watt per steradian	—	W/sr
Specific heat	Joule per kilogram-kelvin	—	$J/kg{\cdot}K$
Stress	Pascal	Pa	N/m^2
Thermal conductivity	Watt per meter-kelvin	—	$W/m{\cdot}K$
Velocity	Meter per second	—	m/s
Viscosity, dynamic	Pascal-second	—	$Pa{\cdot}s$
Viscosity, kinematic	Square meter per second	—	m^2/s
Voltage	Volt	V	W/A
Volume	Cubic meter	—	m^3
Wavenumber	Reciprocal meter	—	(wave)/m
Work	Joule	J	$N{\cdot}m$

always written in singular form. For example:

50 newtons or 50 N

25 millimeters or 25 mm

3.6   Punctuation. Whenever a numerical value is less than one, a zero should precede the decimal point. Periods are not used after any SI unit symbol, except at the end of a sentence. English speaking countries use a dot for the decimal point, others use a comma. Use spaces instead of commas for grouping numbers into threes (thousands). For example:

6 357 831.376 88

not 6,357,831.367,88

3.7   Derived units. The product of two or more units in symbolic form is preferably indicated by a dot midway in relation to unit symbol height. The dot may be dispensed with when there is no risk of confusion with another unit symbol. For example:

Use N · m or N m, but not mN

A solidus (oblique stroke, /) a horizontal line, or negative powers may be used to express a derived unit formed from two others by division. For example:

$$m/s, \frac{m}{s}, \text{ or } m \cdot s^{-1}$$

Only one solidus should be used in a combination of units unless parentheses are used to avoid ambiguity.

3.8   Representation of SI units in systems with limited character sets. For computer printers and other systems which do not have the characters available to print SI units correctly, the methods shown in ISO 2955, Information Processing — Representation of SI and Other Units for Use in Symbols with Limited Character Sets, is recommended.

SECTION 4 — NON-SI UNITS

4.1   Certain units outside the SI are recognized by ISO because of their practical importance in specialized fields. These include units for temperature, time, and angle. Also included are names for some multiples of units such as "liter" (L)* for volume, "hectare" (ha) for land measure and "metric tone" (t) for mass.

4.2   Temperature. The SI base unit for thermodynamic temperature is kelvin (K). Because of the wide usage of the degree Celsius, particularly in engineering and nonscientific areas, the Celsius scale (formerly called the centigrade scale) may be used when expressing temperature. The Celsius scale is related directly to the kelvin scale as follows:

one degree Celsius (1°C) equals one kelvin (1 K), exactly

A Celsius temperature (t) is related to a kelvin temperature (T), as follows:

$t = T - 273.15$

4.3   Time. The SI unit for time is the second. This unit is preferred and should be used when technical calculations are involved. In other cases use of the minute (min), hour (h), day (d), etc. is permissible.

4.4   Angles. The SI unit for plane angle is the radian. The use of arc degrees (°) and its decimal or minute ('), second (") submultiples is permissible when the radian is not a convenient unit. Solid angles should be expressed in steradians.

SECTION 5 — PREFERRED UNITS AND CONVERSION FACTORS

5.1   Preferred units for expressing physical quantities commonly encountered in agricultural engineering work are listed in Table 2. These are presented as an aid to selecting proper units for given applications and to promote consistency when interpretation of the general rules of SI may not produce consistent results. Factors for conversion from old units to SI units are included in Table 2.

SECTION 6 — CONVERSION TECHNIQUES

6.1   Conversion of quantities between systems of units involves careful determination of the number of significant digits to be retained. To convert "1 quart of oil" to "0.946 352 9 liter of oil" is, of course, unrealistic because the intended accuracy of the value does not warrant expressing the conversion in this fashion.

---

*   The International symbol for liter is either the lowercase "l" or the uppercase "L". ASAE recommends the use of uppercase "L" to avoid confusion with the numeral "1".

Table 2
PREFERRED UNITS FOR EXPRESSING PHYSICAL QUANTITIES

Quantity	Application	From: old units	To: SI units	Multiply by:
Acceleration, angular	General	rad/s²	rad/s²	
Acceleration, linear	Vehicle	(mile/h)/s	(km/h)/s	1.609 344[a]
	General (includes acceleration of gravity)[b]	ft/s²	m/s²	0.304 8[a]
Angle, plane	Rotational calculations	r (revolution)	r (revolution)	
		rad	rad	
	Geometric and general	° (deg)	° (decimalized)	
		' (min)	,	1/60[a]
		' (min)	° (decimalized)	
		" (sec)	"	1/3600[a]
		" (sec)		
Angle, solid	Illumination calculations	sr	sr	
Area	Cargo platforms, roof and floor area, frontal areas, fabrics, general	in.²	m²	0.000 645 16[a]
		ft²	m²	0.092 903 04[a]
	Pipe, conduit	in.²	mm²	645.16[a]
		in.²	cm²	6.451 6[a]
		ft²	m²	0.092 903 04[a]
	Small areas, orifices, cross section area of structural shapes	in.²	mm²	645.16[a]
	Brake and clutch contact area, glass, radiators, feed opening	in.²	cm²	6.451 6[a]
	Land, pond, lake, reservoir, open water channel (small)	ft²	m²	0.092 903 04[a]
	(large)	acre	ha	0.404 687 3(d)
	(very large)	mile²	km²	2.589 998
Area per time	Field operations	acre/h	ha/h	0.404 687 3
	Auger sweeps, silo unloader	ft²/s	m²/s	0.092 903 04[a]
Bending moment	(See moment of force)			
Capacitance, electric	Capacitors	μF	μF	
Capacity, electric	Battery rating	A·h	A·h	
Capacity, heat	General	Btu/°f	kJ/K[c]	1.899 101
Capacity, heat, specific	General	Btu/(lb·°F)	kJ/(kg·K)[c]	4.186 8[a]

Table 2 (continued)
PREFERRED UNITS FOR EXPRESSING PHYSICAL QUANTITIES

Quantity	Application	From: old units	To: SI units	Multiply by:
Capacity, volume	(See volume)			
Coefficient of heat transfer	General	Btu/(h·ft²·°F)	W/(m²·K)^c	5.678 263
Coefficient of linear expansion	Shrink fit, general	°F⁻¹, (1/°F)	K⁻¹, (1/K)^c	1.8[a]
Conductance, electric	General	mho	S	1[a]
Conductance, thermal	(See coefficient of heat transfer)			
Conductivity, electric	Material property	mho/ft	S/m	3.280 840
Conductivity, thermal	General	Btu·ft/(h·ft²·°F)	W/(m·K)^c	1.730 735
Consumption, fuel	Off highway vehicles (see also efficiency, fuel)	gal/h	L/h	3.785 412
Consumption, oil	Vehicle performance testing	qt/(1000 miles)	L/(1000 km)	0.588 036 4
Consumption, specific, oil	Engine testing	lb/(hp·h)	g/(kW·h)	608.277 4
		lb/(hp·h)	g/MJ	168.965 9
Current, electric	General	A	A	
Density, current	General	A/in.²	kA/m²	1.550 003
		A/ft²	A/m²	10.763 91
Density, magnetic flux	General	Kilogauss	T	0.1[a]
Density, (mass)	Solid, general; agricultural products, soil, building materials	lb/yd³	kg/m³	0.593 276 3
		lb/in.³	kg/m³	27 679.90
	Liquid	lb/ft³	kg/m³	16.018 46
	Gas	lb/gal	kg/L	0.119 826 4
		lb/ft³	kg/m³	16.018 46
	Solution concentration	—	g/m³, mg/L	—
Density of heat flow rate	Irradiance, general	Btu/(h·ft²)	W/m²	3.154 591[d]
Consumption, fuel	(See flow, volume)			
Consumption, specific fuel	(See efficiency, fuel)			
Drag	(See force)			
Economy, fuel	(See efficiency, fuel)			

Quantity	Description			
Efficiency, fuel	Highway vehicles			
	Economy	mile/gal	km/L	0.415 143 7
	Consumption	—	L/(100 km)	[c]
	Specific fuel consumption	lb/(hp·h)	g/MJ	168.965 9
	Off-highway vehicles			
	Economy	hp·h/gal	kW·h/L	0.196 993 1
	Specific fuel consumption	lb/(hp·h)	g/MJ	168.965 9
	Specific fuel consumption	lb/(hp·h)[f]	kg/(kW·h)[f]	0.608 277 4
Energy, work, enthalpy, quantity of heat	Impact strength	ft-lbf	J	1.355 818
	Heat	Btu	kJ	1.055 056
		kcal	kJ	4.186 8[a]
	Energy, usage, electrical	kW·h	kW·h	3.6
	Mechanical, hydraulic, general	kW·h	MJ	
		ft-lbf	J	1.355 818
		ftˑpdl	J	0.042 140 11
		hp·h	MJ	2.684 520
		hp·h	kW·h	0.745 699 9
Energy per area	Solar radiation	Btu/ft²	MJ/m²	0.011 356 528
Energy, specific	General	cal/g[g]	J/g	4.186 8[a]
		Btu/lb	kJ/kg	2.326[a]
Enthalpy	(See energy)			
Entropy	(See capacity, heat)			
Entropy, specific	(See capacity, heat, specific)			
Floor loading	(See mass per area)			
Flow, heat, rate	(See power)			
Flow, mass, rate	Gas, liquid	lb/min	kg/min	0.453 592 4
		lb/s	kg/s	0.453 592 4
	Dust flow	g/min	g/min	
	Machine work capacity, harvesting, materials handling	ton (short)/h	t/h, Mg/h[a]	0.907 184 7
Flow, volume	Air, gas, general	ft³/s	m³/s	0.028 316 85
		ft³/s	m³/min	1.699 011
	Liquid flow, general	gal/s (gps)	L/s	3.785 412
		gal/s (gps)	m³/s	0.003 785 412
		gal/min (gpm)	L/min	3.785 412
	Seal and packing leakage, sprayer flow	oz/s	mL/s	29.573 53
		oz/min	mL/min	29.573 53

Table 2 (continued)
PREFERRED UNITS FOR EXPRESSING PHYSICAL QUANTITIES

Quantity	Application	From: old units	To: SI units	Multiply by:
	Fuel consumption	gal/h	L/h	3.785 412
	Pump capacity, coolant flow, oil flow	gal/min (gpm)	L/min	3.785 412
	Irrigation sprinkler, small pipe flow	gal/min (gpm)	L/s	0.063 090 20
	River and channel flow	ft³/s	m³/s	0.028 316 85
Flux, luminous	Light bulbs	lm	lm	
Flux, magnetic	Coil rating	maxwell	Wb	0.000 000 01[a]
Force, thrust, drag	Pedal, spring, belt, hand lever, general	lbf	N	4.448 222
		ozf	N	0.278 013 9
		pdl	N	0.138 255 0
		kgf	N	9.806 650
		dyne	N	0.000 01[a]
	Drawbar, breakout, rim pull, winch line pull,[h] general	lbf	kN	0.004 448 222
Force per length	Beam loading	lbf/ft	N/m	14.593 90
	Spring rate	lbf/in.	N/mm	0.175 126 8
Frequency	System, sound and electrical	Mc/s	MHz	1[a]
		kc/s	kHz	1[a]
		Hz, c/s	Hz	1[a]
	Mechanical events, rotational	r/s (rps)	s⁻¹, r/s	1[a]
		r/min (rpm)	min⁻¹, r/min	1[a]
	Engine, power-take-off shaft, gear speed	r/min (rpm)	min⁻¹, r/min	1[a]
	Rotational dynamics	rad/s	rad/s	
Hardness	Conventional hardness numbers, BHN, R, etc. not affected by change to SI			
Heat	(See energy)			
Heat capacity	(See capacity, heat)			
Heat capacity, specific	(See capacity, heat, specific)			
Heat flow rate	(See power)			
Heat flow — density of	(See density of heat flow)			
Heat, specific	General	cal/g·°C	kJ/kg·K	4.186 8[a]
		Btu/lb·°F	kJ/kg·K	4.186 8[a]

Quantity	Application	From	To	Multiply by
Heat transfer coefficient	(See coefficient of heat transfer)			
Illuminance, illumination	General	fc	lx	10.763 91
Impact strength	(See strength, impact)			
Impedance, mechanical	Damping coefficient	lbf·s/ft	N·s/m	14.593 90
Inductance, electric	Filters and chokes, permeance	H	H	1[a]
Intensity, luminous	Light bulbs	candlepower	cd	
Intensity, radiant	General	W/sr	W/sr	
Leakage	(See flow, volume)			
Length	Land distances, maps, odometers	mile	km	1.609 344[a,h]
	Field size, turning circle, braking distance, cargo platforms, rolling circumference, water depth, land leveling (cut and fill)	rod	m	5.029 210[b]
		yd	m	0.914 4
		ft	m	0.304 8[a]
	Row spacing	in.	cm	2.54[a]
	Engineering drawings, product specifications, vehicle dimensions, width of cut, shipping dimensions, digging depth, cross section of lumber, radius of gyration, deflection	in.	mm	25.4[a]
	Precipitation, liquid, daily and seasonal, field drainage (runoff), evaporation and irrigation depth	in.	mm	25.4[a]
	Precipitation, snow depth	in.	cm	2.54[a]
	Coating thickness, filter particle size	mil	µm	25.4[a]
		µin.	µm	0.025 4[a]
		micron	µm	1[a]
	Surface texture			
	Roughness, average	µin.	µm	0.025 4[a]
	Roughness sampling length, waviness height and spacing	in.	mm	25.4[a]
	Radiation wavelengths, optical measurements (interference)	µin.	nm	25.4[a]
Length per time	Precipitation, liquid per hour	in./h	mm/h	25.4[a]
	Precipitation, snow depth per hour	in.h	cm/h	2.54[a]
Load	(See mass)			
Luminance	Brightness	footlambert	cd/m²	3.426 259
Magnetization	Coil field strength	A/in.	A/m	39.370 08

Table 2 (continued)
PREFERRED UNITS FOR EXPRESSING PHYSICAL QUANTITIES

Quantity	Application	From: old units	To: SI units	Multiply by:
Mass	Vehicle mass, axle rating, rated load, tire load, lifting capacity,[i] tipping load, load, quantity of crop, counter mass, body mass	ton (long)	t, Mg[j]	1.016 047
		ton (short)	t, Mg[j]	0.907 184 7
	general	lb	kg	0.453 592 4
		slug	kg	14.593 90
	Small mass	oz	g	28.349 52
Mass per area	Fabric, surface coatings	oz/yd²	g/m²	33.905 75
		lb/ft²	kg/m²	4.882 428
		oz/ft²	g/m²	305.151 7
	Floor loading	lb/ft²	kg/m²	4.882 428
	Application rate, fertilizer, pesticide	lb/acre	kg/ha	1.120 851
	Crop yield, soil erosion	ton (short)/acre	t/ha[j]	2.241 702
Mass per length	General, structural members	lb/ft	kg/m	1.488 164
		lb/yd	kg/m	0.496 054 7
Mass per time	Machine work capacity, harvesting, materials handling	ton (short)/h	t/h, Mg/h[j]	0.907 184 7
Modulus of elasticity	General	lbf/in.²	MPa	0.006 894 757
Modulus of rigidity	(See modulus of elasticity)			
Modulus, section	General	in.³	mm³	16 387.06
		in.³	cm³	16.387 06
Modulus, bulk	System fluid compression	psi	kPa	6.894 757
Moment, bending	(See moment of force)			
Moment of area, second	General	in.⁴	mm⁴	416 231.4
		in.⁴	cm⁴	41.623 14
Moment of force, torque, bending moment	General, engine torque, fasteners, steering torque, gear torque, shaft torque	lbf·in.	N·m	0.112 984 8
		lbf·ft	N·m	1.355 818
		kgf·cm	N·m	0.098 066 5[a]
Moment of inertia, mass	Locks, light torque	ozf·in.	mN·m	7.061 552
Moment of mass	Flywheel, general	lb·ft²	kg·m²	0.042 140 11
Moment of momentum	Unbalance	oz·in.	g·m	0.720 077 8
	(See momentum, angular)			

Quantity	Description	From unit	SI unit	Factor
Moment of section	(See moment of area, second)			
Momentum, linear	General	lbf·ft/s	kg·m/s	0.138 255 0
Momentum, angular	Orsional vibration	lbf·ft²/s	kg·m²/s	0.042 140 11
Permeability	Magnetic core properties	H/ft	H/m	3.280 840
Permeance	(See inductance)			
Potential, electric	General	V	V	
Power	General, light bulbs	Btu/min	W	17.584 17
	Air conditioning, heating	Btu/h	W	0.293 071 1
	Engine, alternator, drawbar, power take-off, hydraulic and pneumatic systems, heat rejection, heat exchanger capacity, water power, electrical power, body heat loss	hp (550 ft·lbf/s)	kW	0.745 699 9
Power per area	Solar radiation	Btu/ft²h	W/m²	3.154 591
Pressure	All pressures except very small	lbf/in.² (psi)	kPa	6.894 757
		in.Hg (60°F)	kPa	3.376 85
		in.H₂O (60°F)	kPa	0.248 84
		mmHg (0°C)	kPa	0.133 322
		kgf/cm²	kPa	98.066 5
		bar	kPa	100.0ᵃ
		lbf/ft²	kPa	0.047 880 26
		atm (normal = 760 torr)	kPa	101.325ᵃ
Pressure, sound level	Very small pressures (high vacuum)	lbf/in.² (psi)	Pa	6 894.757
	Acoustical measurement — when weighting is specified show weighting level in parenthesis following the symbol, for example dB(A)	dB	dB	
Quantity of electricity	General	C	C	
Radiant intensity	(See intensity, radiant)			
Resistance, electric	General	Ω	Ω	
Resistivity, electric	General	Ω·ft	Ω·m	0.304 8ᵃ
		Ω·ft	Ω·cm	30.48ᵃ
Sound pressure level	(See pressure, sound, level)			
Speed	(See velocity)			
Spring rate, linear	(See force per length)			
Spring rate, torsional	General	lbf·ft/deg	N·m/deg	1.355 818

Table 2 (continued)
PREFERRED UNITS FOR EXPRESSING PHYSICAL QUANTITIES

Quantity	Application	From: old units	To: SI units	Multiply by:
Strength, field, electric	General	V/ft	V/m	3.280 840
Strength, field, magnetic	General	Oersted	A/m	79, 577 47
Strength, impact	Materials testing	ft·lbf	J	1.355 818
Stress	General	lbf/in.^{2}	MPa	0.006 894 757
Surface tension	(See tension, surface)			
Temperature	General use	°F	°C	$t_C = (t_F - 32)/1.8$[a]
	Absolute temperature, thermodynamics, gas cycles	°R	K	$T_K = T_{°R}/1.8$[a]
Temperature interval	General use	°F	K^c	$1\ K = 1°C = 1.8°F$[a]
Tension, surface	General	lbf/in.	mN/m	175 126.8
		dyne/cm	mN/m	1[a]
Thermal diffusivity	Heat transfer	ft^2/h	m^2/h	0.092 903 04
Thrust	(See force)			
Time	General	s	s	
		h	h	
		min	min	
	Hydraulic cycle time	s	s	
	Hauling cycle time	min	min	
Torque	(See moment of force)			
Toughness, fracture	Metal properties	ksi·in.$^{0.5}$	MPa·m$^{0.5}$	1.098 843
Vacuum	(See pressure)			
Velocity, angular	(See velocity, rotational)			
Velocity, linear	Vehicle	mile/h	km/h	1.609 344[a]
	Fluid flow, conveyor speed, lift speed, air speed	ft/s	m/s	0.304 8[a]
	Cylinder actuator speed	in./s	mm/s	25.4[a]
	General	ft/s	m/s	0.304 8[a]
		ft/min	m/min	0.304 8[a]
		in./s	mm/s	25.4[a]

Quantity	Application	From unit	To unit	Factor
Velocity, rotational	(See frequency)			
Viscosity, dynamic	General liquids	Centipoise	mPa·s	1[a]
Viscosity, kinematic	General liquids	Centistokes	mm²/s	1[a]
Volume	Truck body, shipping or freight, bucket capacity, earth, gas, lumber, building, general	yd³	m³	0.764 554 9
		ft³	m³	0.028 316 85
	Combine harvester grain tank capacity	Bushel	L	35.239 07
	Automobile luggage capacity	ft³	L	28.316 85
	Gas pump displacement, air compressor, air reservoir, engine displacement			
	Large	in.³	L	0.016 387 06
	Small	in.³	cm³	16.387 06
	Liquid — fuel, lubricant, coolant, liquid wheel ballast	gal	L	3.785 412
		qt	L	0.946 352 9
		pt	L	0.473 176 5
		pt	L	0.473 176 5
	Small quantity liquid	oz	mL	29.573 53
	Irrigation, reservoir	acre·ft	m³	1 233.489[h]
			dam³	1.233 489[h]
	Grain bins	bushel (U.S.)	m³	0.035 239 07
Volume per area	Application rate, pesticide	gal/acre	L/ha	9.353 958
Volume per time	Fuel consumption (also see flow)	gal/h	L/h	3.785 412
Weight	May mean either mass or force — avoid use of weight			
Work	(See energy)			
Young's modulus	(See modulus of elasticity)			

Notes: 1. Quantities are arranged in alphabetical order by principal nouns. For example, surface tension is listed as tension, surface.

2. All possible applications are not listed, but others such as rates can be readily derived. For example, from the preferred units for energy and volume the units for heat energy per unit volume, kJ/m³, may not be derived.

3. Conversion factors are shown to seven significant digits, unless the precision with which the factor is known does not warrant seven digits.

a　Indicates exact conversion factor.

b　Standard acceleration of gravity is 9.806 650 m/s² exactly (Adopted by the General Conference on Weights and Measures).

c　In these expressions, K indicates temperature intervals. Therefore K may be replaced with °C if desired without changing the value or affecting the conversion factor. kJ/(kg·K) = kJ/(kg·°C).

d　Conversions of Btu are based on the International Table Btu.

e　Convenient conversion: 235.215 ÷ (mile per gal) = L/(100 km).

Table 2 (continued)
PREFERRED UNITS FOR EXPRESSING PHYSICAL QUANTITIES

f   ASAE S209 and SAE J708, Agricultural Tractor Test Code, specify kg/(kW·h). It should be noted that there is a trend toward use of g/MJ as specified for highway vehicles.

g   Not to be confused with kcal/g. kcal often called calorie.

h   Official use in surveys and cartography involves the U.S. survey mile based on the U.S. survey foot, which is longer than the international foot by two parts per million. The factors used in this standard for acre, acre foot, rod are based on the U.S. survey foot. Factors for all other old length units are based on the international foot. (See ANSI/ASTM Standard E380-76, Metric Practice).

i   Lift capacity ratings for cranes, hoists, and related components such as ropes, cable chains, etc. should be rated in mass units. Those items such as winches, which can be used for pulling as well as lifting, shall be rated in both force and mass units for safety reasons.

j   The symbol t is used to designate metric ton. The unit metric ton (exactly 1 MG) is in wide use but should be limited to commercial description of vehicle mass, freight mass, and agricultural commodities. No prefix is permitted.

6.2   All conversions, to be logically established, must depend upon an intended precision of the original quantity — either implied by a specific tolerance, or by the nature of the quantity. The first step in conversion is to establish this precision.

6.3   The implied precision of a value should relate to the number of significant digits shown. The implied precision is plus or minus one half unit of the last significant digit in which the value is stated. This is true because it may be assumed to have been rounded from a greater number of digits, and one half of the last significant digit retained is the limit of error resulting from rounding. For example, the number 2.14 may have been rounded from any number between 2.135 and 2.145. Whether rounded or not, a quantity should always be expressed with this implication of precision in mind. For instance, 2.14 in. implies a precision of ±0.005 in., since the last significant digit is in units of 0.01 in.

6.4   Quantities should be expressed in digits which are intended to be significant. The dimension 1.1875 in. may be a very accurate one in which the digit in the fourth place is significant, or it may in some cases be an exact decimalization of a fractional dimension, 1 3/16 in., in which case the dimension is given with too many decimal places relative to its intended precision.

6.5   Quantities should not be expressed with significant zeros omitted. The dimension 2 in. may mean "about 2 in.," or it may, in fact, mean a very accurate expression which should be written 2.0000 in. In the latter case, while the added zeros are not significant in establishing the value, they are very significant in expressing the proper intended precision.

SECTION 7 — RULES FOR ROUNDING

7.1   Where feasible, the rounding of SI equivalents should be in reasonable, convenient, whole units.

7.2   Interchangeability of parts, functionally, physically, or both, is dependent upon the degree of round-off accuracy used in the conversion of the U.S. customary to SI value. American National Standards Institute ANSI/ASTM E380-76, Metric Practice, outlines methods to assure interchangeability.

7.3   Rounding numbers. When a number is to be rounded to fewer decimal places the procedure shall be as follows:

7.3.1   When the first digit discarded is less than 5, the last digit retained shall not be changed. For example, 3.463 25, if rounded to three decimal places, would be 3.463; if rounded to two decimal places, would be 3.46.

7.3.2   When the first digit discarded is greater than 5, or it is a 5 followed by at least one digit other than 0, the last figure retained shall be increased by one unit. For example, 8.376 52, if rounded to three decimal places, would be 8.377; if rounded to two decimal places, would be 8.38.

7.3.3   Round to closest even number when first digit discarded is 5, followed only by zeros.

7.3.4   Numbers are rounded directly to the nearest value having the desired number of decimal places. Rounding must not be done in successive steps to less places. For example:

> 27.46 rounded to a whole number = 27. This is correct because the "0.46" is less than one half. 27.46 rounded to one decimal place is 27.5. This is correct value. But, if the 27.5 is in turn rounded to a whole number, this is successive rounding and the result, 28, is incorrect.

7.4   Inch-millimeter linear dimensioning conversion. 1 inch (in.) = 25.4 millimeters (mm) exactly. The term "exactly" has been used with all exact conversion factors. Conversion factors not so labeled have been rounded in accordance with these rounding procedures. To maintain intended precision during conversion without retaining an unnecessary number of digits, the millimeter equivalent shall be carried to one decimal place more than the inch value being converted and then rounded to the appropriate significant figure in the last decimal place.

CITED STANDARDS

ASAE S209, Agricultural Tractor Test Code
ANSI/ASTM E380-76, Metric Practice
ISO 1000, SI Units and Recommendations for the Use of Their Multiples and of Certain Other Units
ISO 2955, Information Processing — Representation of SI and Other Units for Use in Systems with Limited Character Sets

Index

INDEX